I0038428

Maintenance Engineering

Maintenance Engineering

Emma Affleck

Larsen & Keller
www.larsen-keller.com

Maintenance Engineering
Emma Affleck
ISBN: 978-1-64172-657-3 (Hardback)

© 2022 Larsen & Keller

▤ Larsen & Keller

Published by Larsen and Keller Education,
5 Penn Plaza,
19th Floor,
New York, NY 10001, USA

Cataloging-in-Publication Data

Maintenance engineering / Emma Affleck.
     p. cm.
Includes bibliographical references and index.
ISBN 978-1-64172-657-3
1. Maintainability (Engineering). 2. Maintenance. I. Affleck, Emma.
TA168 .M35 2022
620.004 5--dc23

This book contains information obtained from authentic and highly regarded sources. All chapters are published with permission under the Creative Commons Attribution Share Alike License or equivalent. A wide variety of references are listed. Permissions and sources are indicated; for detailed attributions, please refer to the permissions page. Reasonable efforts have been made to publish reliable data and information, but the authors, editors and publisher cannot assume any responsibility for the validity of all materials or the consequences of their use.

Trademark Notice: All trademarks used herein are the property of their respective owners. The use of any trademark in this text does not vest in the author or publisher any trademark ownership rights in such trademarks, nor does the use of such trademarks imply any affiliation with or endorsement of this book by such owners.

For more information regarding Larsen and Keller Education and its products, please visit the publisher's website www.larsen-keller.com

Table of Contents

Preface

The branch of engineering which focuses on the optimization of procedures, equipment and departmental budgets is known as maintenance engineering. It also focuses on improving the maintainability, availability and reliability of equipment. The primary purpose of maintenance engineering is to ensure that a particular unit is ready for use and maximize its availability while minimizing the costs. Some of the disciplines which contribute knowledge towards maintenance engineering are logistics, probability and statistics. There are numerous applications of this field such as analyzing repetitive equipment failures, forecasting spare parts, estimating repair costs and assessing the requirement for equipment replacements. This book elucidates the concepts and innovative models around prospective developments with respect to maintenance engineering. Some of the diverse topics covered in this book address the varied branches that fall under this category. Scientists and students actively engaged in this field will find this book full of crucial and unexplored concepts.

A short introduction to every chapter is written below to provide an overview of the content of the book:

Chapter 1 - Maintenance is the process of avoiding, analyzing and preventing the degradation of equipment and systems. A few of its types that fall under maintenance engineering are preventive maintenance, risk-based maintenance, corrective maintenance, condition- based maintenance, reliability centered maintenance, etc. This chapter discusses in detail these types of maintenance; **Chapter 2** - The discipline of engineering that aims at achieving better reliability and maintainability of systems with the application of various engineering concepts is termed as maintenance engineering. Total productive maintenance, Deming wheel, kaizen, etc. are different tools used in maintenance engineering. This chapter has been carefully written to provide an easy understanding of maintenance engineering and its tools; **Chapter 3** - Maintenance planning and scheduling are the two major functions that are responsible for the creation of a maintenance program. Some of the maintenance scheduling techniques include six sigma maintenance, lean maintenance, computer aided maintenance, reliability centered maintenance, etc. The topics elaborated in this chapter will help in gaining a better perspective about maintenance planning and scheduling; **Chapter 4** - The analysis and development of strategic models for the improvement of maintenance policies is known as maintenance optimization. It can be categorized into planned maintenance optimization and preventive maintenance optimization. This chapter closely examines maintenance optimization to provide an extensive understanding of the subject; **Chapter 5** - The maintenance activities are planned, organized and monitored by using administrative and technical framework called maintenance management. It is used as a quality process for the management of human errors. This chapter has been carefully written to provide an easy understanding of maintenance management; **Chapter 6** - Various technologies and systems are used for management and optimization of resources such as intelligent maintenance system, predictive management technologies, computerized maintenance management system, etc. This chapter has been carefully written to provide an easy understanding of these maintenance technologies.

Finally, I would like to thank my fellow scholars who gave constructive feedback and my family members who supported me at every step.

Emma Affleck

Introduction to Maintenance

Maintenance is the process of avoiding, analyzing and preventing the degradation of equipment and systems. A few of its types that fall under maintenance engineering are preventive maintenance, risk-based maintenance, corrective maintenance, condition-based maintenance, reliability centered maintenance, etc. This chapter discusses in detail these types of maintenance.

Technical maintenance is a category of maintenance that includes the replacement of unserviceable major parts, assemblies, or subassemblies, and the precision adjustment, testing, and alignment of internal components.

Developing a technical maintenance model to improve operational efficiency is of critical importance for every successful plant or manufacturing unit. Technicians and supervisors shouldn't always be held responsible for economic issues and increased equipment downtimes. The optimization of maintenance management is a complex process, requiring sound field experience and strong analytical skills, combined with best maintenance software.

Optimizing the time of the technical maintenance staff is of critical importance to the overall improvement of maintenance efficacy and efficiency. The implementation of continuous monitoring and analyses of the planned, scheduled and completed maintenance tasks significantly improves equipment availability/costs. The goal of all performed technical maintenance activities should be the plant's reliability. Analyzing the time spent on a maintenance task is more complicated than carrying out proper maintenance efficiency analysis.

Regarding the performance of qualified technicians, the following activities should be analyzed:

- Improvement interventions.
- Qualified (complex) interventions.
- Non-qualified interventions.
- Preparation time.

To optimize the time of qualified technicians, non-qualified tasks can be entirely shifted to less qualified technicians. This will give managers and professional technicians more time to spend on essential areas as:

- Continuous improvement of the preventive maintenance routines.

- FMEA analysis.

- Realization of maintenance instructions.

- Trainings for non-qualified staff.

- Improvement of production equipment.

Analyze Team's Adherence to a Planned Maintenance Program

When frequently and continuously applied, preventive maintenance programs can anticipate and respectively control potential risks. The calendar tool allows managers to easily schedule their preventive maintenance plans. The frequency of the planned maintenance tasks depends on the associated risks as plant reliability, safety or environmental issues. Some interventions can be performed daily, others yearly or in some cases every other year.

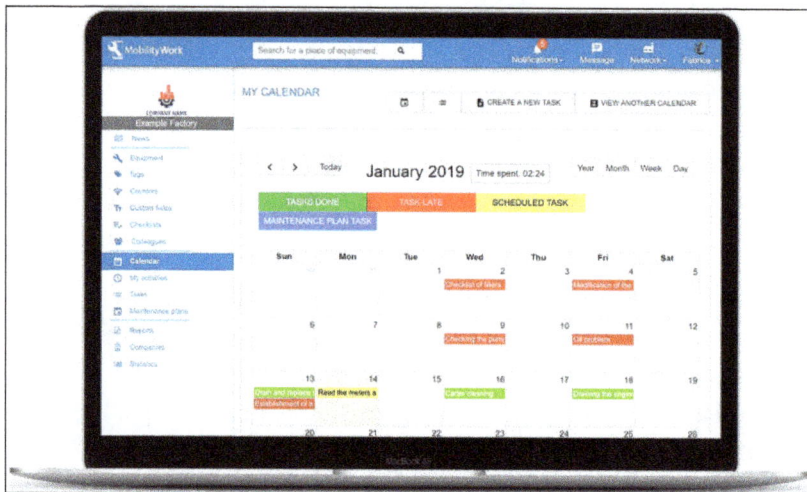

Mobility Work's calendar feature allows you to schedule
all your preventive and predictive maintenance tasks.

Planned and Realized Technical Maintenance Tasks

The main goal of this analysis is to allow rarely performed or neglected preventive maintenance programs to be performed more often. Strictly following the quarterly schedules is much more important than the monthly analysis. However, It is essential to mention that any task performed in a frequency, which is less or equal to a month's duration is critical.

The possible reasons for not respecting the established preventive maintenance regulations can be:

- A surplus of urgent interventions and possible overwork/overtime of the available resources. In this case, quick actions should be undertaken to optimize teams' schedules. Short-term actions as the analysis of value added and non-value-added activities can be implemented. Medium and long-term actions as

FMEA studies to better anticipate failure modes of facilities and evolve the relevance of preventive programs are proven highly efficient.

- The maintenance technician or manager, performing or controlling the maintenance schedule finds its frequency too tight and decides to space it out. In this case, the frequency of the scheduled tasks should be analyzed again and changed accordingly.

Analysis of the Relevance of Curative and Corrective Maintenance

The goal of this analysis is to identify, with managers and technicians, which curative and corrective tasks could have been avoided through:

- Better preventive maintenance schedule.

- Technical changes or investments.

- Better supervision of facilities.

Technical Maintenance Efficiency

Being efficient in technical maintenance means performing all scheduled tasks with the least waste of time possible. In order to define and analyze the value added maintenance tasks, the maintenance workers should be accompanied for one or more full days. By doing so, the following steps can be undertaken:

- Measure the rates of the value added and non-value added tasks.

- Identify possible actions to increase value added tasks.

It is very challenging to identify the value added and the non-value added activities.

The following activities give some examples for value added tasks:

- The time spent by the maintenance technician with his manager in order to understand the interventions to be achieved.

- The preparation of tools, industrial supplies and equipment necessary for the planned intervention.

- Switching from one task to another.

- The time spent performing the task.

- The time spent writing andediting maintenance reports.

Examples for non-value added tasks:

- The time spent waiting for the instructions of the day.

- The time spent waiting between two interventions.

- Trips to the industrial supply stores and tools shops.

- The time spent repairing or cleaning a tool or a machine's component that should have been in perfect working order.

The data analysis can quickly analyze all these values and identify the improvement areas to easily optimize the time of a maintenance technician or manager. The improvement areas are identified by analyzing the distribution of the value added and non-value added activities and the causes for the latter ones.

In general there are two main improvement areas:

- Those related to the non-value added actions of a technician.

- Those related to the non-value added activities during task performance.

Quantification of Improvement Areas

There are three possible methods: linear quantification, statistical quantification and quantification based on the technician's opinion. The third method is the only one possible, when measuring the rates of value added and non-value added activities. If considered that an entire maintenance department rarely exceeds 85% of value added activities, an average rate of 65% can be accepted. 6 months can be set as a goal to reach the 85%.

Scheduled Technical Maintenance Analysis

The goal of this analysis is to measure the time efficiency of already performed planned maintenance activities. It is important to mention that some maintenance tasks are not schedulable. However, the urgent curative ones for example, which by definition cannot be planned, represent just a small part of the technician's time at a well-functioning plant. The main task of a good maintenance is to prevent failures and not to fix breakdowns.

Maintenance Planning Analysis

The formula is:

Planning rate = Total hours available for scheduled tasks/Total hours available.

The ideal maintenance planning rate is 100%.

Measure Maintenance Plan Completion Rate

The plan completion rate measures everything that disrupts the daily maintenance schedule.

The formula is:

Completion rate = Total hours for planned and realized tasks/Total hours for scheduled tasks.

Optimize Staffing Expenditure

The effective management of staff costs can turn into potential gain depending on several conditions:

- If these gains are properly used to effectively reduce internal numbers.

- If these gains are the result of the implementation of improvement actions related to the development of a stable maintenance process. On the contrary to the efficiency benefits, the relevance gains represent a potential that is not always feasible in the short term.

The evaluation process involves three steps:

- Step 1: Synthesis of the performed analyses:

	Results	Performance		
		Low	Medium	High
Value added actions		< 65%	65 à 85%	> 85%
Proactive maintenance supervision		< 10%	10 à 30%	> 30%
Planning rate		< 50%	50 à 80%	> 80%
Adherence to preventive programs		< 80%	80 à 95%	> 95%
Curative and corrective tasks		< 70%	30 à 70%	> 30%
Preventable curative tasks		< 50%	30 à 50%	> 30%
Time spent by a qualified technician on qualified tasks and improvements		< 30%	30 à 60%	> 60%

- Step 2: Efficiency Related Gains - the difference between the average rates of the value added activities and 85% (the basis) is considered as the potential gain.

- Step 3: Relevance Related Gains - even though only preventive maintenance activities are relevant, curative and corrective actions cannot be eliminated. Furthermore, preventable curative and corrective maintenance tasks help to define the potential gains coming from relevant maintenance activities.

Indeed, the percentage of all preventable tasks is the gross percentage of the reduced curative and corrective maintenance tasks. There are only three ways to avoid preventable maintenance tasks: through intensive preventive maintenance, through technical improvements and better facilities management and operation.

The chart below contains some examples for indicators, to measure the relevance and efficiency of your technical maintenance department.

Indicators	Units	Formulas	Goals
Preventive maintenance rate.	%	Total preventive maintenance hours/Total maintenance hours.	To measure the progress of maintenance relevance.
Number of urgent calls to "technicians on call".	Number	Number of calls in a given period.	To give an idea about the plant's emergency state and the level of the generated stress.
Completed preventive maintenance programs.	%	Number of completed programs/Number of scheduled programs.	To anticipate future deterioration of reliability.

MTTR (mean time to repair)	Hours	The average time between failure and repair.	To look for solutions to reduce downtimes.
Maintenance planning rate	%	Total hours for scheduled tasks/Total hours available.	To measure and encourage the planning of maintenance activities.
Completion rate	%	Total hours for scheduled and completed tasks/Total hours available.	To measure and encourage the adherence to maintenance planning.
Capacity utilisation rate	%	Total hours for a task/Total hours available.	To verify the proper use of all available resources and the relevance of the calls to an external company.
Waiting list tasks	Days	Total hours of planned tasks awaiting completion/Available daily hours for the maintenance staff.	To eliminate bottlenecks in the work order system.

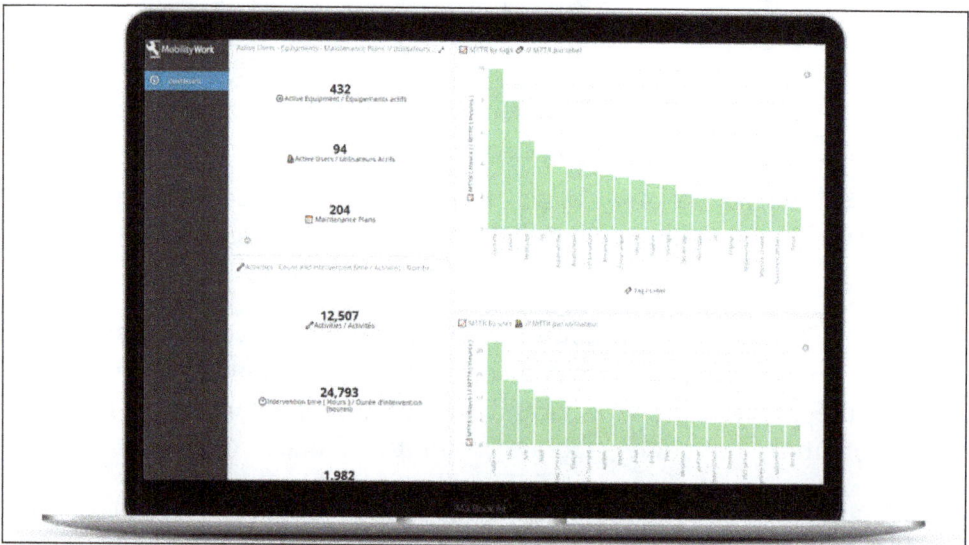

CMMS offers every technical maintenance supervisor all necessary features and tools to efficiently analyze teams' performance. Based on the results important actions can be undertaken to optimize technicians' time and improve the overall plant reliability. By taking a closer look at the team's adherence rate to a planned technical maintenance program, critical problems can be detected and properly solved.

Recognizing the right reasons for not respecting the established preventive maintenance regulations helps managers to better understand their teams' challenges and to improve the distribution of value added and non-value added tasks.

Total Productive Maintenance

Total Productive Maintenance (TPM) is a maintenance program which involves a newly defined concept for maintaining plants and equipment. The goal of the TPM program is to markedly increase production while, at the same time, increasing employee morale and job satisfaction.

TPM brings maintenance into focus as a necessary and vitally important part of the business. It is no longer regarded as a non-profit activity. Down time for maintenance is scheduled as a part of the manufacturing day and, in some cases, as an integral part of the manufacturing process. The goal is to hold emergency and unscheduled maintenance to a minimum.

TPM was introduced to achieve the following objectives. The important ones are listed below:

- Avoid wastage in a quickly changing economic environment.

- Producing goods without reducing product quality.

- Reduce cost.

- Produce a low batch quantity at the earliest possible time.

- Goods send to the customers must be non-defective.

Similarities and Differences between TQM and TPM

The TPM program closely resembles the popular Total Quality Management (TQM) program. Many of the tools such as employee empowerment, benchmarking, documentation, etc. used in TQM are used to implement and optimize TPM. Following are the similarities between the two:

- Total commitment to the program by upper level management is required in both programmes.

- Employees must be empowered to initiate corrective action.

- A long range outlook must be accepted as TPM may take a year or more to implement and is an on-going process. Changes in employee mind-set toward their job responsibilities must take place as well.

Table: The difference between TQM and TPM.

Category	TQM	TPM
Object	Quality (output and effects)	Equipment (Input and cause)
Mains of attaining goal	Systematize the management. It is software oriented	Employees participation and it is hardware oriented
Target	Quality for PPM	Elimination of losses and wastes.

TPM Targets

- P:

    ○ Obtain Minimum 80% OPE.

    ○ Obtain Minimum 90% OEE (Overall Equipment Effectiveness).

    ○ Run the machines even during lunch. (Lunch is for operators and not for machines).

- Q: Operate in a manner, so that there are no customer complaints.

- C: Reduce the manufacturing cost by 30%.

- D: Achieve 100% success in delivering the goods as required by the customer.

- S: Maintain an accident free environment.

- M: Increase the suggestions by 3 times. Develop Multi-skilled and flexible workers.

Motives of TPM	• Adoption of life cycle approach for improving the overall performance of production equipment.
	• Improving productivity by highly motivated workers which is achieved by job enlargement.
	• The use of voluntary small group activities for identifying the cause of failure, possible plant and equipment modifications.
Uniqueness of TPM	The major difference between TPM and other concepts is that the operators are also made to involve in the maintenance process. The concept of "I (Production operators) Operate, You (Maintenance department) fix" is not followed.
TPM Objectives	• Achieve zero defects, zero breakdown and zero accidents in all functional areas of the organization.
	• Involve people in all levels of organization.
	• Form different teams to reduce defects and Self Maintenance.

Direct benefits of TPM	• Increase productivity and OPE (over all plant efficiency) by 1.5 or 2 times.
	• Rectify customer complaints.
	• Reduce the manufacturing cost by 30%.
	• Satisfy the customer needs by 100 % (Delivering the right quantity at the right time, in the required quality).
	• Reduce accidents.
	• Follow pollution control measures.
Indirect benefits of TPM	• Higher confidence level among the employees.
	• Keep the work place clean, neat and attractive.
	• Favorable change in the attitude of the operators.
	• Achieve goals by working as team.
	• Horizontal deployment of a new concept in all areas of the organization.
	• Share knowledge and experience.
	• The workers get a feeling of owning the machine.

Overall Equipment Efficiency

$$OEE = A \times PE \times Q$$

- A - Availability of the machine: Availability is proportion of time machine is actually available out of time it should be available.

  $A = (MTBF - MTTR)/MTBF$.

  MTBF - Mean Time Between Failures = (Total Running Time)/Number of Failures.

  MTTR - Mean Time To Repair.

- PE - Performance Efficiency: It is given by $RE \times SE$:

  ○ Rate efficiency (RE): Actual average cycle time is slower than design cycle time because of jams, etc. Output is reduced because of jams.

  ○ Speed efficiency (SE): Actual cycle time is slower than design cycle time machine output is reduced because it is running at reduced speed.

- Q - Refers to quality rate: Which is percentage of good parts out of total produced sometimes called "yield".

Steps in Introduction of TPM in an Organization

Step A - Preparatory Stage

- Step 1 - Announcement by Management to all about TPM introduction in the organization: Proper understanding, commitment and active involvement of the top management in needed for this step. Senior management should have awareness programmes, after which announcement is made to all. Publish it in the house magazine and put it in the notice board. Send a letter to all concerned individuals if required.

- Step 2 - Initial education and propaganda for TPM: Training is to be done based on the need. Some need intensive training and some just an awareness. Take people who matters to places where TPM already successfully implemented.

- Step 3 - Setting up TPM and departmental committees: TPM includes improvement, autonomous maintenance, quality maintenance etc., as part of it. When committees are set up it should take care of all those needs.

- Step 4 - Establishing the TPM working system and target: Now each area is benchmarked and fix up a target for achievement.

- Step 5 - A master plan for institutionalizing: Next step is implementation leading to institutionalizing wherein TPM becomes an organizational culture. Achieving PM award is the proof of reaching a satisfactory level.

Step B - Introduction Stage

Suppliers as they should know that we want quality supply from them. Related companies and affiliated companies who can be our customers, sisters concerns etc. Some may learn from us and some can help us and customers will get the communication from us that we care for quality output.

Stage C - Implementation

In this stage eight activities are carried which are called eight pillars in the development of TPM activity.

Of these four activities are for establishing the system for production efficiency, one for initial control system of new products and equipment, one for improving the efficiency of administration and are for control of safety, sanitation as working environment.

Stage D - Institutionalising Stage

By all these activities one would has reached maturity stage. Now is the time for applying for PM award. Also think of challenging level to which you can take this movement.

Organization Structure for TPM Implementation

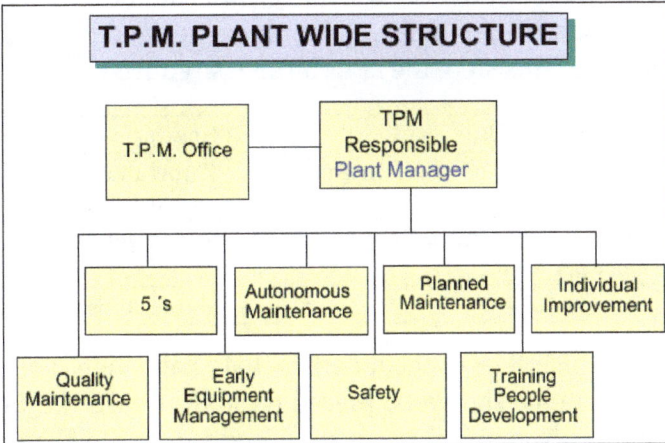

T.P.M. PLANT WIDE STRUCTURE

Pillars of TPM

Pillar 1 - 5S

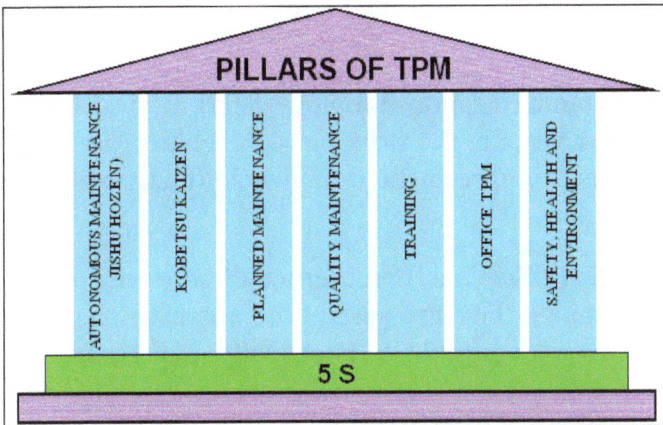

TPM starts with 5S. Problems cannot be clearly seen when the work place is unorganized. Cleaning and organizing the workplace helps the team to uncover problems. Making problems visible is the first step of improvement.

Japanese Term	English Translation	Equivalent 'S' term
Seiri	Organisation	Sort
Seiton	Tidiness	Systematise
Seiso	Cleaning	Sweep
Seiketsu	Standardisation	Standardise
Shitsuke	Discipline	Self - Discipline

1. SEIRI – Sort out: This means sorting and organizing the items as critical, important, frequently used items, useless, or items that are not need as of now.

Unwanted items can be salvaged. Critical items should be kept for use nearby and items that are not be used in near future, should be stored in some place. For this step, the worth of the item should be decided based on utility and not cost. As a result of this step, the search time is reduced.

Priority	Frequency of Use	How to use
Low	Less than once per year, Once per year	Throw away, Store away from the workplace
Average	At least 2/6 months, Once per month, Once per week	Store together but offline
High	Once Per Day	Locate at the workplace

2. SEITON – Organise: The concept here is that "Each items has a place, and only one place". The items should be placed back after usage at the same place. To identify items easily, name plates and colored tags has to be used. Vertical racks can be used for this purpose, and heavy items occupy the bottom position in the racks.

3. SEISO – Shine the workplace: This involves cleaning the work place free of burrs, grease, oil, waste, scrap etc. No loosely hanging wires or oil leakage from machines.

4. SEIKETSU – Standardization: Employees has to discuss together and decide on standards for keeping the work place/Machines/pathways neat and clean. These standards are implemented for whole organization and are tested/Inspected randomly.

5. SHITSUKE – Self-discipline: Considering 5S as a way of life and bring about self-discipline among the employees of the organization. This includes wearing badges, following work procedures, punctuality, dedication to the organization etc.

Pillar 2 - Autonomous Maintenance

This pillar is geared towards developing operators to be able to take care of small maintenance tasks, thus freeing up the skilled maintenance people to spend time on more value added activity and technical repairs. The operators are responsible for upkeep of their equipment to prevent it from deteriorating.

Policy

1. Uninterrupted operation of equipments.

2. Flexible operators to operate and maintain other equipments.

3. Eliminating the defects at source through active employee participation.

4. Stepwise implementation of JH activities.

Jishu Hozen Targets

1. Prevent the occurrence of 1A/1B because of JH.

2. Reduce oil consumption by 50%.

3. Reduce process time by 50%.

4. Increase use of JH by 50%.

Steps in Jishu Hozen

1. Preparation of employees.

2. Initial cleanup of machines.

3. Take counter measures.

4. Fix tentative JH standards.

5. General inspection.

6. Autonomous inspection.

7. Standardization.

8. Autonomous management.

Each of the above mentioned steps is discussed in detail:

1. Train the Employees: Educate the employees about TPM, its advantages, JH advantages and Steps in JH. Educate the employees about abnormalities in equipments.

2. Initial cleanup of machines:

    ◦ Supervisor and technician should discuss and set a date for implementing step 1.

    ◦ Arrange all items needed for cleaning.

    ◦ On the arranged date, employees should clean the equipment completely with the help of maintenance department.

    ◦ Dust, stains, oils and grease has to be removed.

    ◦ Following are the thing that has to be taken care while cleaning. They are Oil leakage, loose wires, unfastened nits and bolts and worn out parts.

    ◦ After clean up problems are categorized and suitably tagged. White tag is place where problems can be solved by operators. Pink tag is placed where the aid of maintenance department is needed.

- ° Content of tag is transferred to a register.

- ° Make note of area which were inaccessible.

- ° Finally close the open parts of the machine and run the machine.

3. Counter Measures:

- ° Inaccessible regions had to be reached easily. E.g. If there are many screw to open a fly wheel door, hinge door can be used. Instead of opening a door for inspecting the machine, acrylic sheets can be used.

- ° To prevent work out of machine parts necessary action must be taken.

- ° Machine parts should be modified to prevent accumulation of dirt and dust.

4. Tentative Standard:

- ° JH schedule has to be made and followed strictly.

- ° Schedule should be made regarding cleaning, inspection and lubrication and it also should include details like when, what and how.

5. General Inspection:

- ° The employees are trained in disciplines like Pneumatics, electrical, hydraulics, lubricant and coolant, drives, bolts, nuts and Safety.

- ° This is necessary to improve the technical skills of employees and to use inspection manuals correctly.

- ° After acquiring this new knowledge the employees should share this with others.

- ° By acquiring this new technical knowledge, the operators are now well aware of machine parts.

6. Autonomous Inspection:

- ° New methods of cleaning and lubricating are used.

- ° Each employee prepares his own autonomous chart/schedule in consultation with supervisor.

- ° Parts which have never given any problem or part which don't need any inspection are removed from list permanently based on experience.

- ° Including good quality machine parts. This avoid defects due to poor JH.

- ° Inspection that is made in preventive maintenance is included in JH.

- ° The frequency of cleanup and inspection is reduced based on experience.

7. Standardization:

- Upto the previous stem only the machinery/equipment was the concentration. However in this step the surroundings of machinery are organized. Necessary items should be organized, such that there is no searching and searching time is reduced.

- Work environment is modified such that there is no difficulty in getting any item.

- Everybody should follow the work instructions strictly.

- Necessary spares for equipment is planned and procured.

8. Autonomous Management:

- OEE and OPE and other TPM targets must be achieved by continuous improve through Kaizen.

- PDCA (Plan, Do, Check and Act) cycle must be implemented for Kaizen.

Pillar 3 - Kaizen

Kaizen is for small improvements, but carried out on a continual basis and involve all people in the organization. Kaizen is opposite to big spectacular innovations. Kaizen requires no or little investment. The principle behind is that "a very large number of small improvements are more effective in an organizational environment than a few improvements of large value. This pillar is aimed at reducing losses in the workplace that affect our efficiencies. By using a detailed and thorough procedure we eliminate losses in a systematic method using various Kaizen tools. These activities are not limited to production areas and can be implemented in administrative areas as well.

Kaizen Policy

1. Practice concepts of zero losses in every sphere of activity.

2. Relentless pursuit to achieve cost reduction targets in all resources.

3. Relentless pursuit to improve over all plant equipment effectiveness.

4. Extensive use of PM analysis as a tool for eliminating losses.

5. Focus of easy handling of operators.

Kaizen Target

Achieve and sustain zero loses with respect to minor stops, measurement and adjustments, defects and unavoidable downtimes. It also aims to achieve 30% manufacturing cost reduction.

Tools used in Kaizen

1. PM analysis.

2. Why - Why analysis.

3. Summary of losses.

4. Kaizen register.

5. Kaizen summary sheet.

The objective of TPM is maximization of equipment effectiveness. TPM aims at maximization of machine utilization and not merely machine availability maximization. As one of the pillars of TPM activities, Kaizen pursues efficient equipment, operator and material and energy utilization that is extremes of productivity and aims at achieving substantial effects. Kaizen activities try to thoroughly eliminate 16 major losses.

Table: 16 Major losses in an organisation.

Loss	Category
1. Failure losses - Breakdown loss 2. Setup/adjustment losses 3. Cutting blade loss 4. Startup loss 5. Minor stoppage/Idling loss 6. Speed loss - operating at low speeds 7. Defect/rework loss 8. Scheduled downtime loss	Losses that impede equipment efficiency.
9. Management loss 10. Operating motion loss 11. Line organization loss 12. Logistic loss 13. Measurement and adjustment loss	Loses that impede human work efficiency.
14. Energy loss 15. Die, jig and tool breakage loss 16. Yield loss	Loses that impede effective use of production resources.

Table: Classification of losses.

Aspect	Sporadic Loss	Chronic Loss
Causation	Causes for this failure can be easily traced. Cause-effect relationship is simple to trace.	This loss cannot be easily identified and solved. Even if various counter measures are applied.

Remedy	Easy to establish a remedial measure.	These types of losses are caused because of hidden defects in machine, equipment and methods.
Impact/Loss	A single loss can be costly.	A single cause is rare - a combination of causes trends to be a rule.
Frequency of occurrence	The frequency of occurrence is low and occasional.	The frequency of loss is more.
Corrective action	Usually the line personnel in the production can attend to this problem.	Specialists in process engineering, quality assurance and maintenance people are required.

Pillar 4 - Planned Maintenance

It is aimed to have trouble free machines and equipment producing defect free products for total customer satisfaction. This breaks maintenance down into 4 "families" or group which was defined earlier:

1. Preventive Maintenance.

2. Breakdown Maintenance.

3. Corrective Maintenance.

4. Maintenance Prevention.

With Planned Maintenance we evolve our efforts from a reactive to a proactive method and use trained maintenance staff to help train the operators to better maintain their equipment.

Policy

1. Achieve and sustain availability of machines.

2. Optimum maintenance cost.

3. Reduces spares inventory.

4. Improve reliability and maintainability of machines.

Target

1. Zero equipment failure and break down.

2. Improve reliability and maintainability by 50%.

3. Reduce maintenance cost by 20%.

4. Ensure availability of spares all the time.

Six Steps in Planned Maintenance

1. Equipment evaluation and recoding present status.

2. Restore deterioration and improve weakness.

3. Building up information management system.

4. Prepare time based information system, select equipment, parts and members and map out plan.

5. Prepare predictive maintenance system by introducing equipment diagnostic techniques and

6. Evaluation of planned maintenance.

Pillar 5 - Quality Maintenance

It is aimed towards customer delight through highest quality through defect free manufacturing. Focus is on eliminating non-conformances in a systematic manner, much like Focused Improvement. We gain understanding of what parts of the equipment affect product quality and begin to eliminate current quality concerns, and then move to potential quality concerns. Transition is from reactive to proactive (quality control to quality assurance).

QM activities are to set equipment conditions that preclude quality defects, based on the basic concept of maintaining perfect equipment to maintain perfect quality of products. The condition are checked and measure in time series to very that measure values are within standard values to prevent defects. The transition of measured values is watched to predict possibilities of defects occurring and to take counter measures beforehand.

Policy

1. Defect free conditions and control of equipments.

2. QM activities to support quality assurance.

3. Focus of prevention of defects at source.

4. Focus on poka-yoke.

5. In-line detection and segregation of defects.

6. Effective implementation of operator quality assurance.

Target

1. Achieve and sustain customer complaints at zero.

2. Reduce in-process defects by 50%.

3. Reduce cost of quality by 50%.

Data Requirements

Quality defects are classified as customer end defects and in house defects. For customer-end data, we have to get data on:

1. Customer end line rejection.

2. Field complaints.

In-house, data include data related to products and data related to process.

Data Related to Product

1. Product wise defects.

2. Severity of the defect and its contribution - major/minor.

3. Location of the defect with reference to the layout.

4. Magnitude and frequency of its occurrence at each stage of measurement.

5. Occurrence trend in beginning and the end of each production/process/changes. (Like pattern change, ladle/furnace lining etc.)

6. Occurrence trend with respect to restoration of breakdown/modifications/periodical replacement of quality components.

Data Related to Processes

1. The operating condition for individual sub-process related to men, method, material and machine.

2. The standard settings/conditions of the sub-process.

3. The actual record of the settings/conditions during the defect occurrence.

Pillar 6 - Training

It is aimed to have multi-skilled revitalized employees whose morale is high and who has eager to come to work and perform all required functions effectively and independently. Education is given to operators to upgrade their skill. It is not sufficient know only "Know-how" by they should also learn "Know-why". By experience they gain, "Know-how" to overcome a problem what to be done. This they do without knowing the root cause of the problem and why they are doing so. Hence it become necessary to train them on knowing "Know-why". The employees should be trained to achieve the four phases of skill. The goal is to create a factory full of experts. The different phase of skills are:

1. Phase 1: Do not know.

2. Phase 2: Know the theory but cannot do.

3. Phase 3: Can do but cannot teach.

4. Phase 4: Can do and also teach.

Policy

1. Focus on improvement of knowledge, skills and techniques.

2. Creating a training environment for self-learning based on felt needs.

3. Training curriculum/tools/assessment etc conductive to employee revitalization.

4. Training to remove employee fatigue and make work enjoyable.

Target

1. Achieve and sustain downtime due to want men at zero on critical machines.

2. Achieve and sustain zero losses due to lack of knowledge/skills/techniques.

3. Aim for 100 % participation in suggestion scheme.

Steps in Educating and Training Activities

1. Setting policies and priorities and checking present status of education and training.

2. Establish of training system for operation and maintenance skill up gradation.

3. Training the employees for upgrading the operation and maintenance skills.

4. Preparation of training calendar.

5. Kick-off of the system for training.

6. Evaluation of activities and study of future approach.

Pillar 7 - Office TPM

Office TPM should be started after activating four other pillars of TPM (JH, KK, QM, PM). Office TPM must be followed to improve productivity, efficiency in the administrative functions and identify and eliminate losses. This includes analyzing processes and procedures towards increased office automation. Office TPM addresses twelve major losses. They are:

1. Processing loss.

2. Cost loss including in areas such as procurement, accounts, marketing, sales leading to high inventories.

3. Communication loss.

4. Idle loss.

5. Set-up loss.

6. Accuracy loss.

7. Office equipment breakdown.

8. Communication channel breakdown, telephone and fax lines.

9. Time spent on retrieval of information.

10. Non availability of correct on line stock status.

11. Customer complaints due to logistics.

12. Expenses on emergency dispatches/purchases.

Start Office TPM

A senior person from one of the support functions e.g. Head of Finance, MIS, Purchase etc. should be heading the sub-committee. Members representing all support functions and people from Production and Quality should be included in subcommittee. TPM co-ordinate plans and guides the subcommittee:

1. Providing awareness about office TPM to all support departments.

2. Helping them to identify P, Q, C, D, S, M in each function in relation to plant performance.

3. Identify the scope for improvement in each function.

4. Collect relevant data.

5. Help them to solve problems in their circles.

6. Make up an activity board where progress is monitored on both sides - results and actions along with Kaizens.

7. Fan out to cover all employees and circles in all functions.

Kobetsu Kaizen Topics for Office TPM

- Inventory reduction.

- Lead time reduction of critical processes.

- Motion and space losses.

- Retrieval time reduction.

- Equalizing the work load.

- Improving the office efficiency by eliminating the time loss on retrieval of information, by achieving zero breakdown of office equipment like telephone and fax lines.

Office TPM and its Benefits

- Involvement of all people in support functions for focusing on better plant performance.

- Better utilized work area.

- Reduce repetitive work.

- Reduced inventory levels in all parts of the supply chain.

- Reduced administrative costs.

- Reduced inventory carrying cost.

- Reduction in number of files.

- Reduction of overhead costs (to include cost of non-production/non capital equipment).

- Productivity of people in support functions.

- Reduction in breakdown of office equipment.

- Reduction of customer complaints due to logistics.

- Reduction in expenses due to emergency dispatches/purchases.

- Reduced manpower.

- Clean and pleasant work environment.

P Q C D S M in Office TPM

- P - Production output lost due to want of material, Manpower productivity, Production output lost due to want of tools.

- Q - Mistakes in preparation of cheques, bills, invoices, payroll, customer returns/warranty attributable to BOPs, Rejection/rework in BOP's/job work, office area rework.

- C - Buying cost/unit produced, Cost of logistics - inbound/outbound, cost of carrying inventory, cost of communication, demurrage costs.

- D - Logistics losses (delay in loading/unloading):

  ◦ Delay in delivery due to any of the support functions.

  ◦ Delay in payments to suppliers.

  ◦ Delay in information.

- S - Safety in material handling/stores/logistics, safety of soft and hard data.

- M - Number of kaizens in office areas.

How Office TPM Supports Plant TPM

Office TPM supports the plant, initially in doing Jishu Hozen of the machines (after getting training of Jishu Hozen), as in Jishu Hozen at the:

- Initial stages machines are more and manpower is less, so the help of commercial departments can be taken, for this.

- Office TPM can eliminate the lodes on line for no material and logistics.

Extension of Office TPM to Suppliers and Distributors

This is essential, but only after we have done as much as possible internally. With suppliers it will lead to on-time delivery, improved 'in-coming' quality and cost reduction. With distributors it will lead to accurate demand generation, improved secondary distribution and reduction in damages during storage and handling. In any case we will have to teach them based on our experience and practice and highlight gaps in the system which affect both sides. In case of some of the larger companies, they have started to support clusters of suppliers.

Pillar 8 - Safety, Health and Environment

Target

- Zero accident,

- Zero health damage,

- Zero fires.

In this area focus is on to create a safe workplace and a surrounding area that is not damaged by our process or procedures. This pillar will play an active role in each of the other pillars on a regular basis.

A committee is constituted for this pillar which comprises representative of officers as well as workers. The committee is headed by Senior vice President (Technical). Utmost importance to Safety is given in the plant. Manager (Safety) is looking after functions related to safety. To create awareness among employees various competitions like safety slogans, quiz, drama, posters, etc. related to safety can be organized at regular intervals.

Preventive Maintenance

Preventive maintenance attempts to prevent any probable failures/breakdowns resulting in production stoppages. It is said that Preventive maintenance is a stich in time that saves time. So it follows a slogan that "prevention is better than cure".

Preventive maintenance refers to maintenance action performed to keep or retain a machine/equipment or asset in a satisfactory operating condition through periodic inspections, lubrication, calibration, replacements and overhauls.

Preventive Maintenance Involves

1. Periodic inspection of equipment/machinery to uncover condition that leads to production breakdown and harmful depreciation. Upkeeps of plant machinery to correct such conditions while they are still in a minor stage.

2. The key to all good preventive maintenance programmes, however is inspection.

3. Regular cleaning, greasing and oiling of moving parts.

4. Replacement of worn out parts before they fail to operate.

5. Periodic overhauling of the entire machine.

6. Machines or equipment's which are liable to sudden failures should be installed in duplicate e.g. motors, pumps, transformers and compressors etc.

Features of Preventive Maintenance

A well-conceived preventive maintenance programme should possess the following features:

1. Proper identification of all items to be included in the maintenance programme.

2. Adequate records covering, volume of work, associated costs etc.

3. Inspection with a definite schedule with standing order on specific assignments.

4. Use of checklists by inspectors.

5. An inspection frequency schedule.

6. A crew of well qualified inspectors with competency of simple repairs, as and when small trouble is noticed.

7. Administrative procedures which provide necessary fulfilment as well as follow up on programme.

Objectives of Preventive Maintenance

1. To minimize the possibility of unanticipated production interruption or major breakdown by uncovering any condition which may lead to it.

2. To make plant, equipment and machinery always available and ready for use.

3. To maintain the value of equipment and machinery by periodic inspections, repairs, overhauls etc.

4. To reduce the work content of maintenance jobs.

5. To maintain the optimum productive efficiency of the plant equipment and machinery.

6. To maintain the operational accuracy of the plant equipment.

7. To achieve maximum production at minimum repair cost.

8. To ensure safety of life and limbs of the workmen along with plant equipment and machines etc.

9. To maintain the operational ability of the plant as a whole.

Procedure of Preventive Maintenance

There is no readymade, on the shelf, preventive maintenance procedure for any industry or enterprise involved in manufacturing activities. In view of the fact that all industries differ in size, location, layout, construction, resources, machinery and its age so as to suit the requirements of an individual industrial plant, the preventive maintenance programmes are specifically framed.

A well-conceived preventive maintenance programme has the following elements, features or steps to be adhered to in general:

1. Who should perform preventive maintenance?

2. Where to start preventive maintenance?

3. What to inspect regarding preventive maintenance?

4. What to inspect for?

5. What should be the frequency of Inspection?

6. When to inspect or inspection schedules?

7. What are preventive maintenance stages?

8. Training of Maintenance staff.

9. Motivation Techniques.

10. Maintenance of Records of Preventive Maintenance.

11. Material Management for Maintenance.

12. Control and Evaluation of Preventive maintenance.

In view of the elements of PM mentioned above for establishing a sound preventive maintenance system in a manufacturing enterprise, we require extra manpower, maintenance facilities, testing equipment's and spare parts etc. to start with but in the long run it provides a lot of benefits by way of reduction in production losses, down time and repair costs etc.

Thus the essential requirements for a sound preventive maintenance can be listed as follows:

1. Proper identification of machines/equipment's and tools: Every item must be uniquely identified by a prominent serial/identity number.

2. Adequate past records must be available for all equipment's being utilized. It should furnish complete details regarding previous maintenance operations/activities.

3. Breakdown/Failures Data: Sufficient breakdown information regarding criticality and frequency of failures must be available for all machines. This would be needed for the purpose of failure identification, failure diagnostics, analysis as well as final rectification.

4. Secondary data: In fact it is a sort of experienced data for similar equipment being utilized.

5. Manufacturer's utilization recommendations: Regarding the use of a particular machine i.e. how to utilize and provide P.M.

6. Service manuals, instruction and maintenance sheets.

7. Consumables and replacable parts/components should be available as and when needed.

8. Availability of requisite skilled manpower may be engineers, inspectors and technicians.

9. Availability/provision of test rigs/equipment's i.e., test rigs, sensors etc.

10. Clear instructions with a check list regarding preventive and corrective measures must be available to ensure proper functioning of the system.

11. Users feedback and cooperation: The user of the equipment/machine must provide feedback to the manufacturer regarding actual functioning of the equipment.

12. Management Support: For establishing a preventive maintenance system, the commitment of top management is very essential for the implementation of preventive maintenance policy of the organization.

Applications of Preventive Maintenance

1. This system of maintenance is applicable for automated or continuous production process e.g., steel mills, chemical plants and automobile industries.

2. In some of the abovementioned practical situations the cost of lost production due to a failure/breakdown can be extremely high. Besides, such heavy cost of failures, the breakdowns may be totally destructive in nature i.e., the failure of a small equipment may lead to complete breakdown of the system. Hence preventive maintenance system is essential in such situations to ensure continuous and failure free plant operation.

3. In the failure of equipment's such as boilers, turbines, pressure vessels and lifting devices the results may be fatal sometimes. Thus in order to avoid any loss of human life and health hazards, a proper preventive maintenance system must be adopted.

4. Some common examples where preventive maintenance is adopted are as follows:

   - P.M. of machine tools,
   - P.M. of pressure vessels or boilers,
   - P.M. of steam and gas turbines,
   - P.M. of heat exchangers,
   - P.M. of mobile compressors and generators,
   - Overhead cranes,
   - Small power plants,
   - Elevators,
   - Vehicles.

Preventive maintenance is subdivided into following two categories:

1. Running

2. Shut down

Running maintenance means that maintenance work carried out even when machine or equipment is in service, while shut down maintenance is concerned with maintenance work carried out only when the machine/equipment is not in operation.

Advantages of Preventive Maintenance

1. Reduction in breakdown time and associated breakdown elements.

2. Reduces the odd time repairs and over time to the maintenance staff.

3. Fewer large scale and repetitive repairs.

4. Less member of standby equipment and spare parts required.

5. Greater safety to work force/employees due to reduced breakdowns.

6. Increased life of equipment and machines.

7. The work load of the maintenance staff can be properly planned.

8. It improves the availability of facilities.

9. Optimum production efficiency can be achieved by employing preventive maintenance.

10. Maintenance and repair cost reduce heavily.

11. It improves the quality of product and reduces rejections.

12. Production cost goes down by adopting RM.

13. Regular planned servicing and adjustment maintains and provides a high level of plant output, better equipment performance and better product quality.

14. Healthy, hygienic, safe and an accident free work environment can be achieved with the application of scientific preventive maintenance. This would promote industrial relations since workers do not lose any type of incentive due to breakdowns or accidents.

15. Reduction in inventory of spare parts.

Limitations of Preventive Maintenance

1. When the cost of failure prevention is always greater than cost of failure rectification the process of P.M. is very costly e.g., batch production-bridge construction.

2. The type of maintenance requires extra facilities and lead to under/poor utilization of basic facilities for RM.

3. For small scale manufacturing units which are mainly undertaking job and batch production, the P.M system is not suited and economically justified.

Risk-based Maintenance

Risk-based maintenance (RBM) prioritizes maintenance resources toward assets that carry the most risk if they were to fail. It is a methodology for determining the most economical use of maintenance resources. This is done so that the maintenance effort across a facility is optimized to minimize any risk of a failure.

A risk-based maintenance strategy is based on two main phases:

1. Risk assessment.

2. Maintenance planning based on the risk.

The maintenance type and frequency are prioritized based on the risk of failure. Assets that have a greater risk and consequence of failure are maintained and monitored more frequently. Assets that carry a lower risk are subjected to less stringent maintenance programs. Implementing a risk-based maintenance process means that the total risk of failure is minimized across the facility in the most economical way. The monitoring and maintenance programs for high risk assets are typically condition-based maintenance programs.

1. Suitable applications: Risk-based maintenance is a suitable strategy for any maintenance plan. As a methodology, it provides a systematic approach to determine the most appropriate asset maintenance plans. Upon implementation of these maintenance plans, the risk of asset failure will be low.

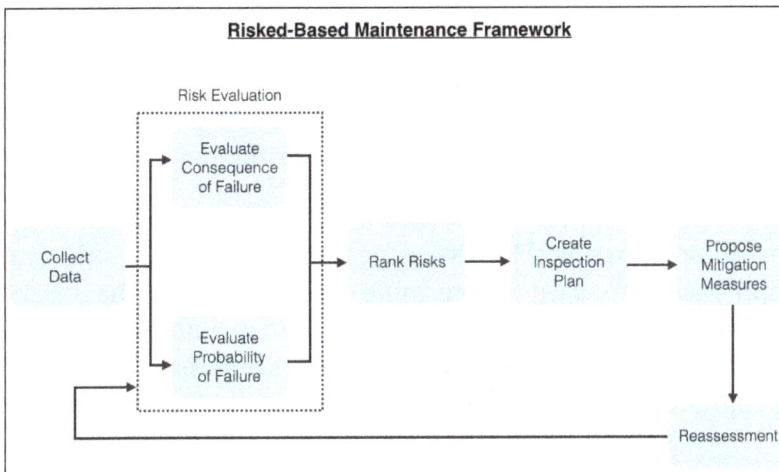

Risked-Based Maintenance Framework

The risk-based maintenance framework is applied to each system in a facility. A system, for example, may be a high-pressure vessel. That system will have neighbouring systems that pass fluid to and from the vessel. The likely failure modes of the system are first determined. Then, a typical risk-based maintenance framework is applied to each risk. The framework is shown in the figure.

2. Collect data: For each identified risk, data needs to be collected. This includes information about the risk, its general consequences and the general methods used to mitigate and predict the risk.

3. Risk evaluation: At the risk evaluation stage, both the probability of the risk and the consequence of the risk are quantified in the context of the facility under consideration.

4. Rank risks: With the risk evaluation complete, the probability and consequence are combined to determine the total risk. This total risk is ranked against pre-determined levels of risk. As a result, the risk is either acceptable or unacceptable.

5. Creating an inspection plan: If the risk is unacceptable, a plan to inspect the system using a condition monitoring approach is determined. Or, if it is more cost appropriate, and technically feasible a preventative maintenance program might be selected.

6. Propose mitigation measure: At this stage the proposal for mitigating the risk, using the condition monitoring and maintenance approach, is prepared.

7. Reassessment: Finally, the proposal is evaluated against other factors – such as legal and regulatory requirements. If the proposal's needs are not met, then the process starts again. Otherwise, the maintenance proposal is put into place.

Assess the Risk

Assessing the risk of failure is one of the most important aspects of risk based maintenance. The more accurately this is done, the better the risk based maintenance outcomes will be.

There is no one standard method for assessing risk. Qualitative, semi-quantitative and quantitative approaches are used to determine the possible risks that exist. To estimate the likelihood of these risks, the methods that are used include deterministic and probabilistic approaches. 62 different approaches to assessing risk are described in Tixier. The most appropriate approach will depend on the data that is available to evaluate each risk.

Corrective Maintenance

Corrective maintenance is a special type of maintenance activity carried out to restore an item/machine when it has failed to meet an acceptable condition. Further it is basically a rectification process which is always adopted after the occurrence of a breakdown.

It may be executed in the form of:

1. Repair may be minor or major.

2. Replacement may be partial or total.

In fact, down time due to breakdown may consist not only of time taken to complete the repair work but also delays caused by lack of resources or information. Repair time is a function of management technique so, engineering techniques and maintainability. Corrective or planned maintenance is required not only when the asset/machine item fails but also when indicated by condition based criteria.

Characteristics of Corrective Maintenance

1. A correction maintenance activity is generally planned.

2. A planned or unplanned corrective maintenance operation depends on the nature of breakdown and type of equipment/machine.

3. The maintenance work is taken up after the occurrence of a breakdown and with some permissible time lag.

4. Breakdown maintenance should not be very serious in nature as far as production losses, down time, loss of human life etc. are concerned.

5. Breakdown of individual equipment should not affect considerably the overall production loss.

In general breakdown maintenances are predictable and expected failures and hence they may be rectified over a long period of time without any time constraints.

Objectives of Corrective Maintenance

1. To get equipment/machine back into operation as quickly as possible in order to minimize the interruption to production. These objectives are directly related with production capacity, costs of production, product quality and consumer satisfaction.

2. To control the cost of the operation of repair shops.

3. To keep the cost of repair crew under control, including regular and overtime of labour costs.

4. To control the investment in replacement of parts/components that are used/ required when machines are repaired.

5. To control the investment required for back up machines. These replace manu- facturing machines are needed until the repairs are completed.

6. To perform the appropriate amount of repairs at each malfunction of the asset/ equipment.

7. To restore an asset in working order.

8. To maintain the operation availability of the plant and infrastructural facili- ties.

9. To avoid any sudden and heavy failure (breakdown) in future.

Typical Causes of Equipment Breakdown

The causes of equipment breakdown may be as follows:

1. Failure to replace worn out components/parts.

2. Lack of lubrication.

3. Neglected cooling arrangement/system.

4. Indifference towards minor faults.

5. External factors such as wrong fuel, too low or too high line voltage etc.

6. Indifference towards equipment vibrations, unusual sounds coming out of the rotating parts and equipment getting too much heated up.

Advantages of Corrective Maintenance

1. Emergency maintenance requirements are reduced.

2. Heavy down time losses are reduced.

3. Plant availability is increased.

4. Results in better utilization of plant facilities.

5. Safety level is improved and hence there are less chances of accidents.

6. Provides sufficient information concerning the maintenance replacement and repair.

Limitations of Breakdown Maintenance

1. Breakdowns generally occur at inappropriate times. It may lead to a poor hurried maintenance and excessive delays in production schedules.

2. It involves prolonged down time due to non-availability of requisite manpower and spare parts, they may lead to overtime practice also.

3. It becomes impossible to plan workload and distribution of maintenance workforce for balanced and proper attention of all equipment's.

4. Reduction in production output.

5. There are increased chances of accidents and less safety for workforce.

6. It leads to faster plant deterioration.

7. Corrective maintenance cannot be employed for those industrial plants/enterprises which are regulated by statutory provisions for example boilers and cranes.

8. The maintenance of product quality is difficult.

9. Loss of direct profits.

Reliability Centred Maintenance

Reliability Centred Maintenance (RCM) is a very powerful methodology which, when properly applied, can drive significant improvements in equipment reliability and plant performance while, at the same time, making sure that the money being spent on Predictive and Preventive maintenance programs is optimised.

One of the most common approaches for developing or improving a Preventive Maintenance program is to use Reliability Centred Maintenance (RCM).

Reliability Centred Maintenance Standards

We are fortunate that there are several standards that define the Reliability Centred Maintenance approach. This includes the SAE standard JA1011: Evaluation Criteria for Reliability-Centered Maintenance (RCM) Processes, and IEC 60300.3.11-2011 Dependability management - Application guide - Reliability Centred Maintenance.

RCM is a structured process which sequentially asks the following seven questions about the asset or system under review:

1. Functions - What are the functions and associated performance standards of the asset in its present operating context?

2. Functional Failures - In what ways does it fail to fulfill its functions?

3. Failure Modes - What causes each functional failure?

4. Failure Effects - What happens when each failure occurs?

5. Failure Consequences - In what way does each failure matter?

6. Proactive Tasks - What can be done to predict or prevent each failure?

7. Default Actions - What should be done if a suitable proactive task cannot be found?

Functions and Performance Standards

A key concept of Reliability Centred Maintenance is the understanding that the primary purpose of a preventive maintenance program is to ensure that equipment continues to do what the business requires it to do, in its present operating context. Therefore, the first step is to ensure that we full understand what it is that we need the equipment to do (its functions), and the level of performance that is required if the business is to meet its objectives.

The Functions of equipment consists of Primary Functions (normally associated with the reasons that the equipment was acquired in the first place), and Secondary Functions (additional requirements, often to do with Safety, Efficiency etc.).

For many equipment items, there can be several (potentially as much as 20 or 30 in some cases) functions. For each of these functions, we must, where possible, quantify the level of performance required. For example, it would not be sufficient to state that the Primary Function of a pump is to pump a liquid from A to B, we would also need to specify the flow rate required, if production output targets were to be met.

As a result, if done properly, answering this first question alone can take between 25% and 35% of the total time required for an entire RCM analysis.

Functional Failures

Functional Failures (or Failed States) simply define, for each function, the states under which the equipment does not fulfil its Functions. For each function, there is a need to consider both complete failure (for example where the equipment fails to operate at all), and partial failures (where the equipment operates, but does not operate at a level sufficient to meet the performance standard associated with that Function.

Failure Modes

Once each Functional Failure has been identified, the next step in the RCM process is to identify all the events which are reasonably likely to cause each failed state. These events are known as Failure Modes. "Reasonably likely" Failure Modes include those which have occurred on the same or similar equipment operating in the same context, failures which are currently being prevented by existing maintenance regimes and failures which have not happened yet but which are considered to be real possibilities in the context in question. The list should include failures caused by human errors (on the part of operators and maintainers) and design flaws so that all reasonably likely causes of equipment failure can be identified and dealt with appropriately. It is also important to identify the cause of each failure in enough detail to ensure that time and efforts are not wasted trying to treat symptoms instead of causes. On the other hand, it is equally important to ensure that time is not wasted on the analysis itself by going into too much detail.

Failure Effects

The fourth step in the RCM process entails listing Failure Effects, which describes what happen when each failure mode occurs. These descriptions should include all the information needed to support the evaluation of the consequences of the failure (in the next step in the process), such as:

- What evidence (if any) that the failure has occurred?

- In what ways (if any) it poses a threat to safety or the environment?

- In what ways (if any) it affects production or operations?

- What physical damage (if any) is caused by the failure?

- What must be done to repair the failure?

Identifying the relevant Failure Modes and Failure Effects is also a comparatively lengthy activity – typically requiring around 30-35% of the time required for the entire RCM analysis.

Failure Consequences

Another key concept underpinning Reliability Centred Maintenance is that the primary objective of a Preventive Maintenance program is not necessarily to avoid or minimise failures themselves, but to avoid or minimise the consequences of those failures. There is little point in spending a lot of time and money preventing failures that have little or no consequences associated with them. On the other hand, if a failure has serious consequences, we may be able to justify going to great lengths to avoid those consequences. In this way, the RCM process focuses attention on the maintenance activities which have most effect on the performance of the organization, and diverts energy away from those which have little or no effect.

The fifth step in the RCM process classifies the consequences associated with each failure mode as belonging to one of the following four groups:

- Hidden failure consequences: Hidden failures have no direct impact, but they expose the organization to multiple failures with serious, often catastrophic, consequences. (Most of these failures are associated with protective devices which are not fail-safe).

- Safety and environmental consequences: A failure has safety consequences if it could hurt or kill someone. It has environmental consequences if it could lead to a breach of any corporate, regional, national or international environmental standard.

- Operational consequences: A failure has operational consequences if it affects production (output, product quality, customer service or operating costs in addition to the direct cost of repair).

- Non-operational consequences: Evident failures which fall into this category affect neither safety nor production, so they involve only the direct cost of repair.

The consequence evaluation process also shifts emphasis away from the idea that all failures are bad and must be prevented. In so doing, it focuses attention on the maintenance activities which have most effect on the performance of the organization, and diverts energy away from those which have little or no effect.

Default Actions

RCM recognizes three major categories of default actions, as follows:

- Failure-finding: Failure-finding tasks entail checking hidden functions periodically to determine whether they have failed (whereas condition-based tasks entail checking if something is failing).

- Redesign: Redesign entails making any one-off change to the built-in capability of a system. This includes modifications to the hardware and also covers once-off changes to procedures.

- No Scheduled Maintenance: As the name implies, this default entails making no effort to anticipate or prevent failure modes to which it is applied, and so those failures are simply allowed to occur and then repaired. This default is also called run-to-failure.

Condition-based Maintenance

Condition-based maintenance (CBM) is a maintenance strategy that monitors the actual condition of an asset to decide what maintenance needs to be done. CBM dictates that maintenance should only be performed when certain indicators show signs of decreasing performance or upcoming failure. Checking a machine for these indicators may include non-invasive measurements, visual inspection, performance data and scheduled tests. Condition data can then be gathered at certain intervals, or continuously (as is done when a machine has internal sensors). Condition-based maintenance can be applied to mission critical and non-mission critical assets.

Unlike in planned maintenance (PM), where maintenance is performed based upon predefined scheduled intervals, condition-based maintenance is performed only after a decrease in the condition of the equipment has been observed. Compared with preventive maintenance, this increases the time between maintenance repairs, because maintenance is done on an as-needed basis.

Goal of Condition-based Maintenance

The goal of condition based maintenance is to monitor and spot upcoming equipment failure so maintenance can be proactively scheduled when it is needed – and not before. Asset conditions need to trigger maintenance within a long enough time period before failure, so work can be finished before the asset fails or performance falls below the optimal level.

Advantages

- CBM is performed while the asset is working, which lessens the chances of disruption to normal operations.

- Reduces the cost of asset failures.

- Improves equipment reliability.

- Minimizes unscheduled downtime due to catastrophic failure.

- Minimizes time spent on maintenance.

- Minimizes overtime costs by scheduling the activities.

- Minimizes requirement for emergency spare parts.

- Optimizes maintenance intervals (more optimal than manufacturer recommendations).

- Improves worker safety.

- Reduces the chances of collateral damage to the system.

Disadvantages

- Condition monitoring test equipment is expensive to install, and databases cost money to analyse.

- Cost to train staff–you need a knowledgeable professional to analyze the data and perform the work.

- Fatigue or uniform wear failures are not easily detected with CBM measurements.

- Condition sensors may not survive in the operating environment.

- May require asset modifications to retrofit the system with sensors.

- Unpredictable maintenance periods.

Example of condition-based maintenance:

Motor vehicles come with a manufacturer-recommended interval for oil replacements. These intervals are based on manufacturers' analysis, years of performance data and experience. However, this interval is based on an average or best guess rather than the actual condition of the oil in any specific vehicle. The idea behind condition based maintenance is to replace the oil only when a replacement is needed, and not on a predetermined schedule.

In the example of industrial equipment, oil analysis can perform an additional function too. By looking at the type, size and shape of the metal particulates that are suspended in the oil, the health of the equipment it is lubricating can also be determined.

Types of Condition based Maintenance

Oil analysis

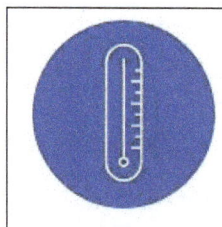

Infrared

Electrical

Operational

Performance

Ultrasonic Acoustic

Vibration Analysis

There are various types of condition-based monitoring techniques. Here are a few common examples:

- Vibration analysis: Rotating equipment such as compressors, pumps and motors all exhibit a certain degree of vibration. As they degrade, or fall out of alignment, the amount of vibration increases. Vibration sensors can be used to detect when this becomes excessive.

- Infrared: IR cameras can be used to detect high-temperature conditions in energized equipment.

- Ultrasonic: Detection of deep subsurface defects such as boat hull corrosion.

- Acoustic: Used to detect gas, liquid or vacuum leaks.

- Oil analysis: Measures the number and size of particles in a sample to determine asset wear.

- Electrical: Motor current readings using clamp on ammeters.

- Operational performance: Sensors throughout a system measure pressure, temperature, flow etc.

Challenges of Condition-based Maintenance

- Condition-based maintenance requires an investment in measuring equipment and staff up-skilling so the initial costs of implementation can be high.

- CBM introduces new maintenance techniques, which can be difficult to implement due to resistance within an organization.

- Older equipment can be difficult to retrofit with sensors and monitoring equipment, or can be difficult to access during production to spot measure.

- With CBM in place, it still requires competence to turn performance information from a system into actionable proactive maintenance items.

Formula for CBM

CBM = Cost Savings + Higher System Reliability

Condition-based maintenance allows preventive and corrective actions to be scheduled at the optimal time, thus reducing the total cost of ownership. Today, improvements in technology are making it easier to gather, store and analyze data for CBM. In particular, CBM is highly effective where safety and reliability is the paramount concern such as the aircraft industry, semiconductor manufacturing, nuclear, oil and gas, et cetera.

Data Collection

Data can be collected from the system by two different methods:

- Spot readings can be performed at regular intervals using portable instruments.

- Sensors can be retrofitted to equipment or installed during manufacture for continuous data collection.

Critical systems that require considerable upfront capital investment, or that could affect the quality of the product that is produced, need up-to-the-minute data collection. More expensive systems have built in intelligence to self-monitor in real time. For example, sensors throughout an aircraft monitor numerous systems while in flight and on the ground to help identify issues before they become life-threatening. Typically, CBM is not used for non-critical systems and spot readings will suffice.

Equipment Maintenance

Equipment maintenance is any process used to keep a business's equipment in reliable working order. It may include routine upkeep as well as corrective repair work. Equipment may include mechanical assets, tools, heavy off-road vehicles, and computer systems. The resources needed to keep it all in good repair will vary by type. For instance, repairs made on heavy construction equipment won't look the same as those performed on automated food processing machines.

Types of Equipment Maintenance Workers

Equipment maintenance workers may include technicians, supervisors, and managers.

- Maintenance technicians: Equipment maintenance technicians handle general upkeep and repair work on a business's equipment. They may also be involved in diagnostic testing and routine inspections as directed by their supervisors. Individual technicians may specialize in working on specific types of equipment, or they might provide general maintenance services.

- Maintenance supervisors: Equipment maintenance supervisors oversee technicians and plan maintenance tasks for each day. They make sure all health and safety requirements are met, manage workloads, and handle preventive maintenance planning.

- Maintenance Managers: An equipment maintenance manager handles high level planning and oversee maintenance supervisors. Their planning is centered around meeting the department's long-term goals rather than day-to-day requirements, though their tasks may overlap somewhat with those of supervisors.

Example of equipment maintenance:

A road construction company owns numerous expensive assets that are vital to their business, including heavy-duty construction equipment. In order to make sure their equipment lasts as long as possible, they perform routine inspections on each asset. The intervals for these types of equipment are often based on hours of usage.

For instance, their motor grader needs certain types of maintenance every 500 hours. These tasks include replacing oil, air, and fuel filters, changing out the hydraulic tank, lubricating bearings and gears, and inspecting fuel tank caps. To keep track of these maintenance requirements, the company logs the number of hours the grader is used each day, and when it reaches a 500 hour interval, they schedule a maintenance inspection.

The company also inspects their power tools after each shift. Their pneumatic jackhammers, for instance, are checked daily for cracks in hoses, abrasion on the bit, and loosening screws. Not only does this help their tools last longer, it also enhances safety by preventing dangerous equipment failures. Their tools also operate more efficiently, helping them remain productive.

Industries that use Equipment Maintenance

Any industry that uses any kind of equipment uses equipment maintenance. Some major examples include the following:

- Food processing: Heavy machinery, mobile equipment, and handheld tools used in food processing all require equipment maintenance.

- Plastics manufacturing: Plastics manufacturing plants use a wide range of heavy and lightweight equipment, all of which needs regular servicing.

- Steel mills: Maintenance workers in steel fabrication plants service equipment ranging from hot rollers and furnaces to portable tools.

- Restaurants: The various types of equipment used in commercial kitchens need to be kept in working order to provide reliable service. Regular equipment maintenance also helps restaurants maintain regulatory compliance.

- Construction: Keeping heavy mobile equipment, handheld power tools, and safety gear in good repair is key to a construction company's efficient operation.

- Automobile manufacturing: Fabricating and assembling vehicle parts requires finely tuned equipment. Production is best when that equipment is kept in good repair.

- Workshops: Workshops use a variety of tools to produce various items, such as woodworking, metal products, and blown glass. Given the exacting standards of this industry, their tools need to be kept in top condition.

References

- Technical-maintenance-time-and-staff-optimization-analysis: mobility-work.com, Retrieved 1 August , 2019

- Tpm-intro: plant-maintenance.com, Retrieved 9 May, 2019

- Preventive-maintenance-meaning-objectives-and-applications, preventive-maintenance, industries: yourarticlelibrary.com, Retrieved 8 August , 2019

- Risk-based-maintenance, maintenance-strategies: fiixsoftware.com, Retrieved 31 March, 2019

- Corrective-maintenance-definition-objectives-and-limitations, maintenance-management: yourarticlelibrary.com, Retrieved 14 July, 2019

- Reliability-improvement/what-is-reliability-centred-maintenance-rcm: assetivity.com.au, Retrieved 17 May, 2019

- Condition-based-maintenance: fiixsoftware.com, Retrieved 19 April, 2019

- Equipment-maintenance, maintenance-applications, learning: onupkeep.com, Retrieved 5 February, 2019

2

Understanding Maintenance Engineering

The discipline of engineering that aims at achieving better reliability and maintainability of systems with the application of various engineering concepts is termed as maintenance engineering. Total productive maintenance, Deming wheel, kaizen, etc. are different tools used in maintenance engineering. This chapter has been carefully written to provide an easy understanding of maintenance engineering and its tools.

Maintenance engineering uses engineering theories and practices to plan and implement routine maintenance of equipment and machinery. This must be done in conjunction with optimizing operating procedures and budgets to attain and sustain the highest levels of reliability and profit. The onslaught of high-tech machinery, multiple infrastructures and systems, and intricate computerized manufacturing and production systems over the past few decades has elevated these jobs to new levels of responsibility and qualification requirements. Maintenance engineers are often required to have knowledge of many types of equipment and machinery.

A person working in this field must have in-depth knowledge of or experience in basic equipment operation, logistics, probability, and statistics. Experience in the operation and maintenance of machinery specific to a company's particular business is also frequently required. Since the position normally requires oral and written communications with various levels of personnel, excellent interpersonal communication and participatory management skills are also desirable.

Maintenance engineering positions require planning and implementing routine and preventive maintenance programs. In addition, regular monitoring of equipment is required to visually detect faults and impending equipment or production failures before they occur. These positions may also require observing and overseeing repairs and maintenance performed by outside vendors and contractors. In a production or manufacturing environment, good maintenance is necessary for smooth and safe daily plant operations. Maintenance engineers not only monitor the existing systems and equipment, they also recommend improved systems and help decide when systems are outdated and in need of replacement. Such a position often involves exchanging ideas and information with other maintenance engineers, production managers, and manufacturing systems engineers. Maintenance engineering not only requires engineers to monitor large production machine operations and heavy duty equipment, but also often requires involvement with computer operations. Maintenance engineers may have to deal with everything from PCs, routers, servers, and software to more complex issues like local and off-site networks, configuration systems, end user support, and scheduled upgrades. Supervision of technical personnel may also be required.

Good maintenance engineering is vital to the success of any manufacturing or processing operation, regardless of size. The maintenance engineer is responsible for the efficiency of daily operations and for discovering and solving any operational problems in the plant. A company's success may depend on a quality maintenance department that can be depended upon to discover systematic flaws and recommend solid, practical solutions. Positions in this field often require a college education in a related field. Although most schools do not offer degrees in maintenance engineering, degrees in mechanical engineering, industrial engineering, or related subjects are preferred.

Maintenance engineers ensure machines and processes in a facility run smoothly. This relies on both preventative and emergency maintenance to keep operations running with as little system downtime as possible. This often involves periodic inspections, commissioning or recommending of system upgrades, troubleshooting equipment and repairing or installing replacement parts. If a problem is found, it is commonly the responsibility of the maintenance engineer to prevent the problem from reoccurring.

Proactive vs Reactive

With the onset of Industry 4.0, manufacturers are moving towards more a proactive approach to maintenance. The sector has been particularly receptive to automated and smart technologies because of their ability to reduce downtime and improve productivity.

A traditional reactive approach to maintenance revolves around waiting for equipment to fail before performing maintenance. The obvious downside to this is that the company is not prepared for unplanned stoppages, which can lead to downtime and the waste of raw materials. For example, if batch production is halted on a food manufacturing line, it can lead to the fouling of ingredients and in some cases damage to the system. The risk of downtime means that maintenance engineers need to make sure equipment is kept functional. As factories have become smarter, so too has maintenance. By using data from sensors on the factory floor, maintenance engineers can plan a schedule and use predictive maintenance to tackle any equipment problems before it causes downtime.

The increasing number of connected sensors in manufacturing facilities means that maintenance is now performed with a data-based approach. Condition monitoring equipment that uses techniques such as vibration analysis and infrared imaging allows the maintenance engineer to perform preventative action.

Looking Forward

In the highly automated manufacturing facilities of the future, machine-to-machine communication, sensor developments, big data and analytics will change the relationship between the factory and the maintenance engineer. Systems could be able to self-diagnose any part failures, order replacement components from a supplier, and install them with minimal human intervention.

More data will be available for maintenance engineers to analyse, which steers staff time away from manual data input and towards high value roles such as updating and monitoring the plant's digital twin — a real-time simulation of the plant in operation.

Maintenance Organization

Maintenance cost can be a significant factor in an organization's profitability. In manufacturing, maintenance cost could consume 2–10% of the company's revenue and may reach up to 24% in the transport industry. So, contemporary management considers maintenance as an integral function in achieving productive operations and high-quality products, while maintaining satisfactory equipment and machines reliability as demanded by the era of automation, flexible manufacturing systems (FMS), "lean manufacturing", and "just-in-time" operations.

However, there is no universally accepted methodology for designing maintenance systems, i.e., no fully structured approach leading to an optimal maintenance system (i.e., organizational structure with a defined hierarchy of authority and span of control; defined maintenance procedures and policies, etc.). Identical product organizations, but different in technology advancement and production size, may apply different maintenance systems and the different systems may run successfully. So, maintenance systems are designed using experience and judgment supported by a number of formal decision tools and techniques. Nevertheless, two vital considerations should be considered: strategy that decides on which level within the plant to perform maintenance, and hence outlining a structure that will support the maintenance; planning that handles day to-day decisions on what maintenance tasks to perform and providing the resources to undertake these tasks.

The maintenance organizing function can be viewed as one of the basic and integral parts of the maintenance management function (MMF). The MMF consists of planning, organizing, implementing and controlling maintenance activities. The management organizes, provides resources (personnel, capital, assets, material and hardware, etc.) and leads to performing tasks and accomplishing targets. Once the plans are created, the management's task is to ensure that they are carried out in an effective and efficient manner. Having a clear mission, strategy, and objectives facilitated by a corporate culture, organizing starts the process of implementation by clarifying job and working relations (chain of command, span of control, delegation of authority).

In designing the maintenance organization there are important determinants that must be considered. The determinants include the capacity of maintenance, centralization vs. decentralization and in-house maintenance *vs* outsourcing. A number of criteria can be used to design the maintenance organization. The criteria include clear roles and responsibilities, effective span of control, facilitation of good supervision and effective reporting, and minimization of costs.

Maintenance managers must have the capabilities to create a division of labor for maintenance tasks to be performed and then coordinate results to achieve a common purpose. Solving performance problems and capitalizing on opportunities could be attained through selection of the right persons, with the appropriate capabilities, supported by continuous training and good incentive schemes, in order to achieve organization success in terms of performance effectiveness and efficiency.

Maintenance Organization Objectives and Responsibility

A maintenance organization and its position in the plant/whole organization is heavily impacted by the following elements or factors:

- Type of business, e.g., whether it is high tech, labor intensive, production or service.

- Objectives may include profit maximization, increasing market share and other social objectives.

- Size and structure of the organization.

- Culture of the organization.

- Range of responsibility assigned to maintenance.

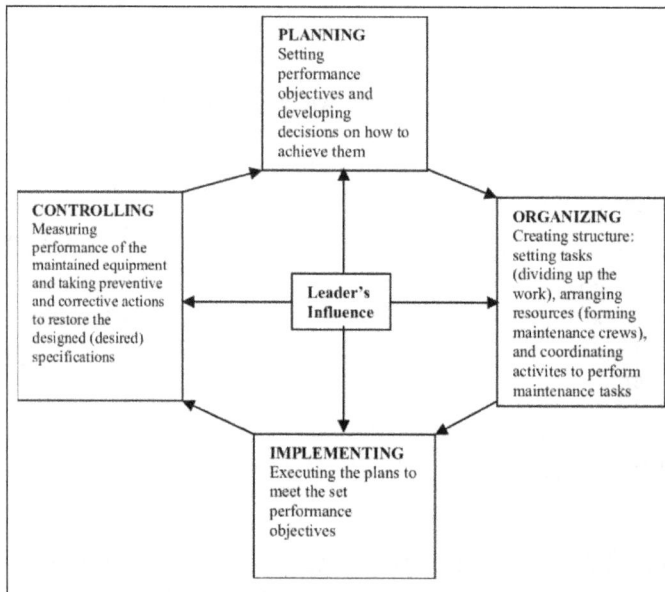

PLANNING
Setting performance objectives and developing decisions on how to achieve them

CONTROLLING
Measuring performance of the maintained equipment and taking preventive and corrective actions to restore the designed (desired) specifications

Leader's Influence

ORGANIZING
Creating structure: setting tasks (dividing up the work), arranging resources (forming maintenance crews), and coordinating activites to perform maintenance tasks

IMPLEMENTING
Executing the plans to meet the set performance objectives

Maintenance organizing as a function of the management process.

Organizations seek one or several of the following objectives: profit maximization, specific quality level of service or products, minimizing costs, safe and clean environment, or human resource development. It is clear that all of these objectives are heavily impacted by maintenance and therefore the objectives of maintenance must be aligned with the objectives of the organization.

The principal responsibility of maintenance is to provide a service to enable an organization to achieve its objectives. The specific responsibilities vary from one organization to another; however they generally include the following according to Duffuaa et al:

- Keeping assets and equipment in good condition, well configured and safe to perform their intended functions.

- Perform all maintenance activities including preventive, predictive; corrective, overhauls, design modification and emergency maintenance in an efficient and effective manner.

- Conserve and control the use of spare parts and material.

- Commission new plants and plant expansions.

- Operate utilities and conserve energy.

Determinants of a Maintenance Organization

The maintenance organization's structure is determined after planning the maintenance capacity. The maintenance capacity is heavily influenced by the level of centralization or decentralization adopted. The issues are: capacity planning, centralization vs decentralization and in-house vs outsourcing.

Maintenance Capacity Planning

Maintenance capacity planning determines the required resources for maintenance including the required crafts, administration, equipment, tools and space to execute the maintenance load efficiently and meet the objectives of the maintenance department. Critical aspects of maintenance capacity are the numbers and skills of craftsmen required to execute the maintenance load. It is difficult to determine the exact number of various types of craftsmen, since the maintenance load is uncertain. Therefore accurate forecasts for the future maintenance work demand are essential for determining the maintenance capacity. In order to have better utilization of manpower, organizations tend to reduce the number of available craftsmen below their expected need. This is likely to result in a backlog of uncompleted maintenance work. This backlog can also be cleared when the maintenance load is less than the capacity. Making long run estimations is one of the areas in maintenance capacity planning that is both critical and not well developed in practice.

Centralization vs Decentralization

The decision to organize maintenance in a centralized, decentralized or a hybrid form depends to a greater extent on the organization is philosophy, maintenance load, size of the plant and skills of craftsmen. The advantages of centralization are:

- Provides more flexibility and improves utilization of resources such highly skilled crafts and special equipment and therefore results in more efficiency.

- Allows more efficient line supervision.

- Allows more effective on the job training.

- Permits the purchasing of modern equipment.

However it has the following disadvantages:

- Less utilization of crafts since more time is required for getting to and from jobs.

- Supervision of crafts becomes more difficult and as such less maintenance control is achieved.

- Less specialization on complex hardware is achieved since different persons work on the same hardware.

- More costs of transportation are incurred due to remoteness of some of the maintenance work.

In a decentralized maintenance organization, departments are assigned to specific areas or units. This tends to reduce the flexibility of the maintenance system as a whole. The range of skills available becomes reduced and manpower utilization is usually less efficient than in a centralized maintenance. In some cases a compromise solution that combines centralization and decentralization is better. This type of hybrid is called a cascade system. The cascade system organizes maintenance in areas and whatever exceeds the capacity of each area is challenged to a centralized unit. In this fashion the advantages of both systems may be reaped.

In-house vs Outsourcing

At this level management considers the sources for building the maintenance capacity. The main sources or options available are in-house by direct hiring, outsourcing, or a combination of in-house and outsourcing. The criteria for selecting sources for building and maintaining maintenance capacity include strategic considerations, technological and economic factors. The following are criteria that can be employed to select among sources for maintenance capacity:

- Availability and dependability of the source on a long term basis.

- Capability of the source to achieve the objectives set for maintenance by the organization and its ability to carry out the maintenance tasks.

- Short term and long term costs.

- Organizational secrecy in some cases may be subjected to leakage.

- Long term impact on maintenance personnel expertise.

- Special agreement by manufacturer or regulatory bodies that set certain specifications for maintenance and environmental emissions.

Examples of maintenance tasks which could be outsourced are:

- Work for which the skill of specialists is required on a routine basis and which is readily available in the market on a competitive basis:

  ○ Installation and periodic inspection and repair of automatic fire sprinkler systems.

  ○ Inspection and repair of air conditioning systems.

  ○ Inspection and repair of heating systems.

  ○ Inspection and repair of main frame computers etc.

- When it is cheaper than recruiting your own staff and accessible at a short notice of time.

Design of the Maintenance Organization

A maintenance organization is subjected to frequent changes due to uncertainty and desire for excellence in maintenance. Maintenance and plant managers are always swinging from supporters of centralized maintenance to decentralized ones, and back again. The result of this frequent change is the creation of responsibility channels and direction of the new organization's accomplishments vs the accomplishments of the former structure. So, the craftsmen have to adjust to the new roles. To establish a maintenance organization an objective method that caters for factors that influence the effectiveness of the organization is needed. Competencies and continuous improvement should be the driving considerations behind an organization's design and re-design.

Current Criteria for Organizational Change

Many organizations were re-designed to fix a perceived problem. This approach in many cases may raise more issues than solve the specific problem. Among the reasons to change a specific maintenance organization's design are:

- Dissatisfaction with maintenance performance by the organization or plant management.

- A desire for increased accountability.

- A desire to minimize manufacturing costs, so maintenance resources are moved to report to a production supervisor, thereby eliminating the (perceived) need for the maintenance supervisor.

- Many plant managers are frustrated that maintenance seems slow paced, that is, every job requires excessive time to get done. Maintenance people fail to understand the business of manufacturing, and don't seem to be part of the team. This failure results in decentralization or distribution of maintenance resources between production units.

- Maintenance costs seem to rise remarkably, so more and more contractors are brought in for larger jobs that used to get done in-house.

Criteria to Assess Organizational Effectiveness

Rather than designing the organization to solve a specific problem, it is more important to establish a set of criteria to identify an effective organization. The following could be considered as the most important criteria:

- Roles and responsibilities are clearly defined and assigned.

- The organization puts maintenance in the right place in the organization.

- Flow of information is both from top-down and bottom-up.

- Span of control is effective and supported with well-trained personal.

- Maintenance work is effectively controlled.

- Continuous improvement is built in the structure.

- Maintenance costs are minimized.

- Motivation and organization culture.

Basic Types of Organizational Models

To provide consistently the capabilities listed above we have to consider three types of organizational designs.

- Entralized maintenance: All crafts and related maintenance functions report to a central maintenance manager. The strengths of this structure are: allows economies of scale; enables in-depth skill development; and enables departments (i.e., a maintenance department) to accomplish their functional goals (not the overall organizational goals). This structure is best suited for small to medium size organizations. The weaknesses of this structure are: it has slow response time to environmental changes; may cause delays in decision making and hence longer response time; leads to poor horizontal coordination among departments and involves a restricted view of organizational goals.

- Decentralized maintenance: All crafts and maintenance craft support staff report to operations or area maintenance. The strengths of this structure are that it allows the organization to achieve adaptability and coordination in production units and efficiency in a centralized overhaul group and it facilitates effective coordination both within and between maintenance and other departments. The weaknesses of this structure are that it has potential for excessive administrative overheads and may lead to conflict between departments.

- Matrix structure, a form of a hybrid structure: Crafts are allocated in some proportion to production units or area maintenance and to a central maintenance function that supports the whole plant or organization. The strengths of this matrix structure are: it allows the organization to achieve coordination necessary to meet dual demands from the environment and flexible sharing of human resources. The weaknesses of this structure are: it causes maintenance employees to experience dual authority which can be frustrating and confusing; it is time consuming and requires frequent meetings and conflict resolution sessions. To remedy the weaknesses of this structure a management with good interpersonal skills and extensive training is required.

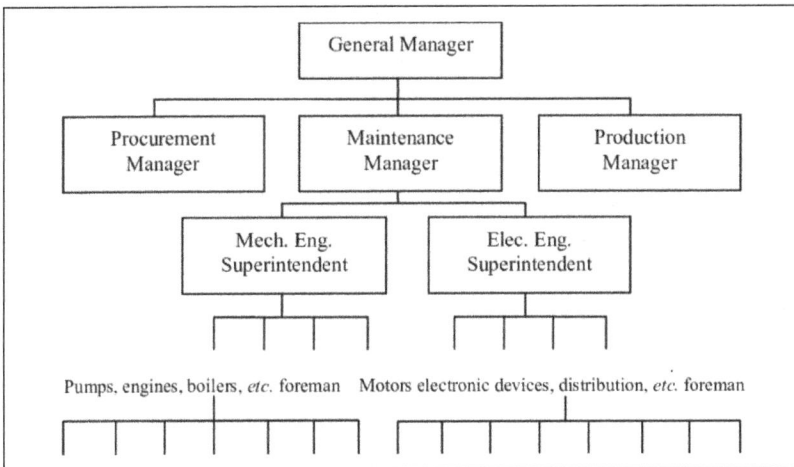

Technicians: Centralized (functional) organizational structure.

Material and Spare Parts Management

The responsibility of this unit is to ensure the availability of material and spare parts in the right quality and quantity at the right time at the minimum cost. In large or medium size organizations this unit may be independent of the maintenance organization; however in many circumstances it is part of maintenance. It is a service that supports the maintenance programs. Its effectiveness depends to a large extent on the standards maintained within the stores system.

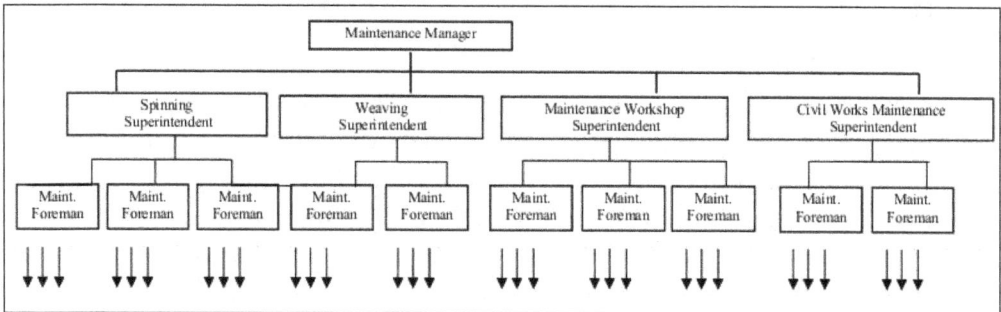

Functionally de-centralized organizational structure of maintenance in a textile factory.

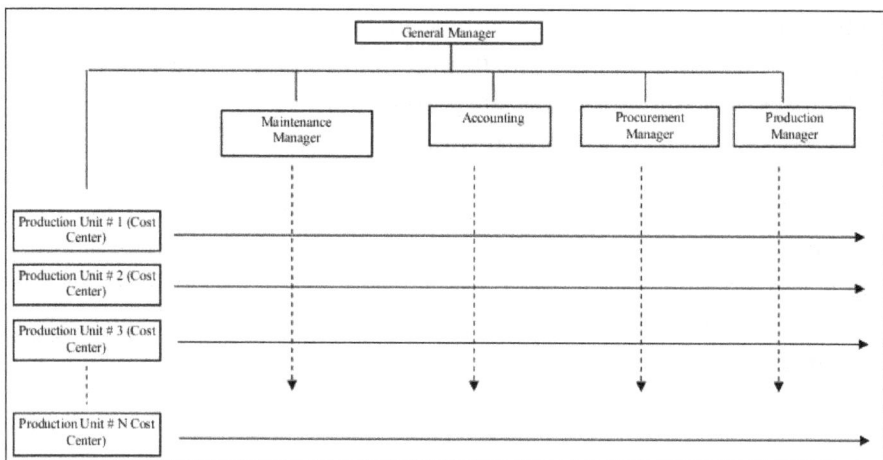

Matrix (de-centralized) organizational structures.

The duties of a material and spare parts unit include:

- Develop in coordination with maintenance effective stocking polices to minimize ordering, holding and shortages costs.

- Coordinate effectively with suppliers to maximize organization benefits.

- Keep good inward, receiving, and safe keeping of all supplies.

- Issue materials and supplies.

- Maintain and update records.

- Keep the stores orderly and clean.

Establishment of Authority and Reporting

Overall administrative control usually rests with the maintenance department, with its head reporting to top management. This responsibility may be delegated within the maintenance establishment. The relationships and responsibility of each maintenance division/section must be clearly specified together with the reporting channels. Each job title must have a job description prescribing the qualifications and the experience needed for the job, in addition to the reporting channels for the job.

Quality of Leadership and Supervision

The organization, procedures, and practices instituted to regulate the maintenance activities and demands in an industrial undertaking are not in themselves a guarantee of satisfactory results. The senior executive and his staff must influence the whole functional activity. Maintenance performance can never rise above the quality of its leadership and supervision. From good leadership stems the teamwork which is the essence of success in any enterprise. Talent and ability must be recognized and fostered; good work must be noticed and commended; and carelessness must be exposed and addressed.

Management and Labor Relations

The success of an undertaking depends significantly on the care taken to form a community of well-informed, keen, and lively people working harmoniously together. Participation creates satisfaction and the necessary team spirit. In modern industry, quality of work life (QWL) programs have been applied with considerable success, in the form of management conferences, work councils, quality circles, and joint conferences identified with the activities. The joint activities help the organization more fully achieve its purposes.

Maintenance Method Tool

Industrial maintenance management can sometimes seem very complicated for many companies. Yet there are a lot of simple tools that can help you master it perfectly. If you want your business to be successful, you need to be well-organized, to communicate smoothly with your collaborators and to make the right choice for your CMMS. TPM, FMECA, Kaizen: all these terms probably sound familiar

Total Productive Maintenance

The TPM method (total productive maintenance) was born in Japan in 1971. It basically aims to progressively change maintenance techniques in order to increase equipment output. This method allows you to avoid unplanned shutdowns, time loss every time a technician turns on a machine or waste when pieces of equipment are deficient. As a result, ineffective equipment or lack of attention from certain technicians are not a problem anymore. In order to get to this point, TPM is divided into three clear tools:

- The overall output rate, which is an indicator for the equipment use rate.

- The 5 Ss:

  ○ Seiri – sort;

  ○ Seiton – set in order;

  ○ Seiso – shine;

  ○ Seiketsu – standardize;

  ○ Shitsuke – sustain.

- Autonomous maintenance, which allows production operators to perform simplified maintenance tasks.

PDCA

The PDCA, or Deming wheel, is a technique aiming to enhance your industrial projects anticipation and management. This tool helps you organize your ideas and divide the work you have to do into several steps in order to ensure that everything goes well. The PDCA acronym corresponds to:

- P – "Plan": plan what you are going to do.

- D – "Do": do what was planned.

- C – "Check": check that the work done corresponds to what was planned in the first place.

- A – "Act": react according to the assessment of the work done.

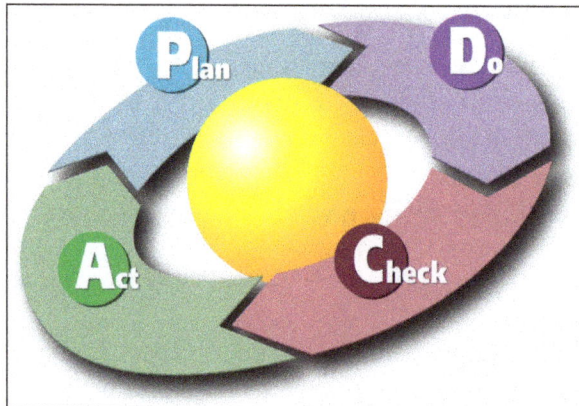

Ishikawa Diagram

The Ishikawa diagram, also known as fishbone diagrams, herringbone diagrams or cause-and-effect diagrams, is used in quality management. It is a way for you to identify the different origins and consequences of a given issue.

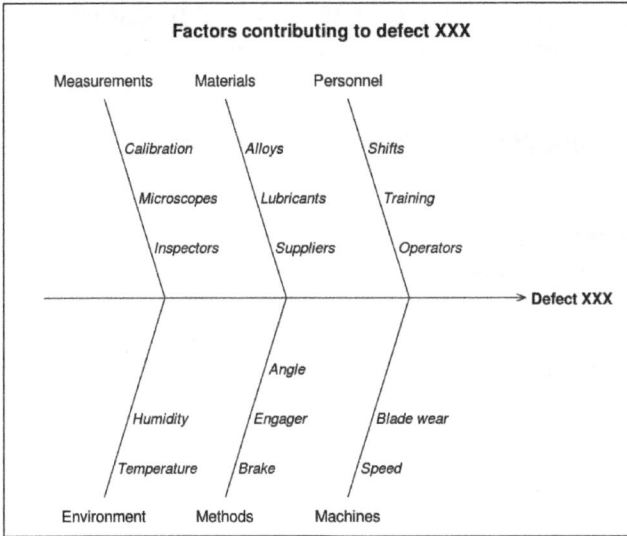

Factors contributing to defect XXX

It is a very simple tool, you just need to list all the possible causes of the problem you encounter and classify them according to different categories. If we take the example of a machine, you just have to create categories such as electrical, mechanical, hydraulic, automation and then list the potential issues. This visual tool is particularly used in risk management as it allows you to anticipate a lot of difficulties that could have dramatic consequences on your company's activity.

The Five Ws

The 5 Ws are questions that you need to ask yourself when you're facing an issue. This is therefore a very simple method to quickly obtain the information you need. The aim is to deal with an industrial problematic and to answer the following questions:

- Who (was involved)?
- What (happened)?
- Where (did it take place)?
- When (did it take place)?
- Why (did it happen)?

Thanks to this method, you can gather a lot of information that will probably help you better understand the issue you are facing. It is then possible to identify spheres of actions, to put them in order of priority and to react with the proper solutions and propositions.

Kaizen

Kaizen is a Japanese method consisting in enhancing a plant's efficiency and output quality thanks to minor but continuous improvement. In order for this method to be perfectly efficient, all employees need to get involved in this continuous improvement process and all have to gather to achieve their objective. To put this method into place, it is necessary to:

- Organize awareness-raising sessions to encourage employees to get involved and to relate;

- Train managers and technicians to this method so that they can completely understand why it is so useful;

- Standardize processes and implement tools such as pdca, presented above, or quality management tools that give everyone the chance to express themselves freely and to share their opinion.

Pareto or ABC analysis

The Pareto analysis, also known as the 20-80 rule, allows you to analyze the major breakdowns in terms of frequency as well as time spent. It states that 20% (or even less) of the causes are responsible for 80% of a plant's problems and helps analyzing all the issues to find an appropriate answer. In order to use this method and to have a general overview of the situation, you first need to take your failure history into consideration. The next-gen CMMS Mobility Work, thanks to its analytics module, displays all your data automatically so that you can use them.

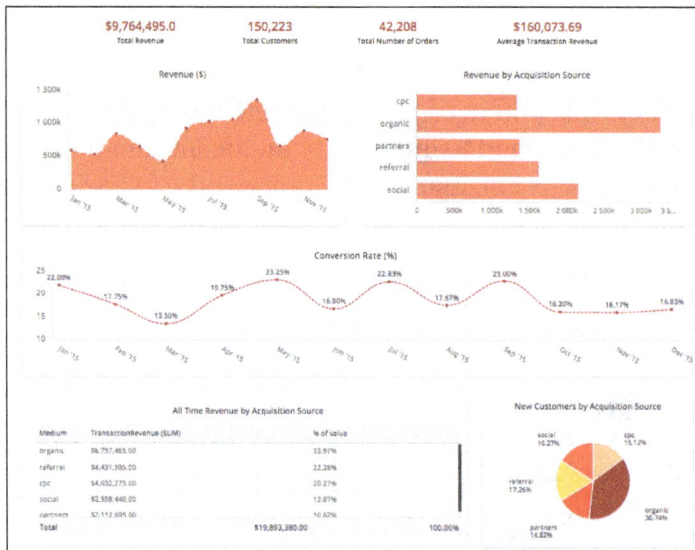

The ABC analysis is based on the same principle as the Pareto analysis as it directly results from it. It basically is a classifying method that is widely used for stock analysis.

Failure Modes, Effects and Criticality Analysis

The failure modes, effects and criticality analysis (FMECA) aim to help you analyze your maintenance interventions as well as your machinery. FMECA allows you to manage your industrial maintenance, as this operational safety tool is also very used in quality approaches.

To make the best use of it, simply follow these few steps:

- Determine the mode of failure as well as the cause.

- Assess the repercussions on the system, the damaged feature as well as the resulting damage.

- Identify the following criteria:

  ◦ N: Number of breakdowns.

  ◦ O: Occurrence.

  ◦ S: Severity.

  ◦ D: Detectability.

- Calculate the criticality thanks to the following formula: Detectability Occurrence.

The FMECA can take different forms (functional, product, process, production mean, flow) and have different effects each time but which finally will allow you to obtain a work document that is absolutely necessary to know what actions to put in place, what interventions to plan, etc.

As a result, this method enhances production results and limits failure problems in the same time, it also allows you to analyze production defects, to constantly look for improvement, etc.

The industrial risk and equipment failure assessment methodology - maintenance method tool:

This risk assessment methodology is particularly used in the maintenance operations planning. In order to make the best use of it, you have to evaluate the failures before carrying out a general risk assessment study.

If you decide to implement this tool, you will be able to identify the type of maintenance that you want to put into place (predictive or preventive for example), to better manage your stocks, to determine the procedures that need to be followed, etc.

All these tools can make your everyday-life easier and, thanks to a next-gen CMMS like Mobility Work, make the industrial maintenance of your company's equipment very smooth. Thanks to our community-based maintenance application, you will be able to communicate easily within your company and to evolve towards 4.0 maintenances.

All these tools won't have any secrets for you and you won't have to struggle with the difficulties of industrial maintenance anymore.

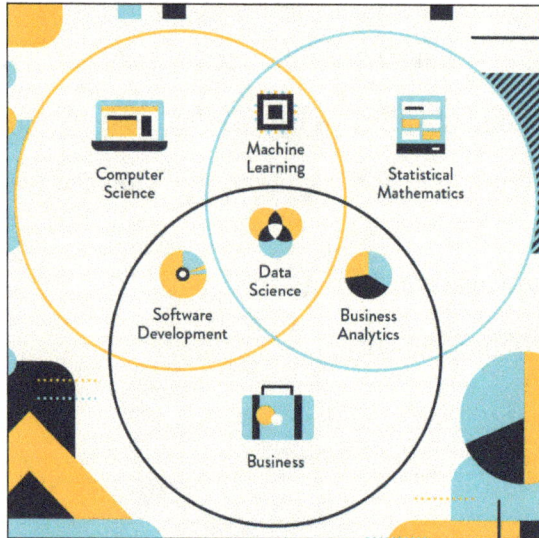

Common Maintenance Problems

Many maintenance departments today "fight fires" instead of approaching their problems systematically. Prevention is a far better goal than trying to solve problems as they arise. While this strategy may be a little costly at first, it is not nearly as expensive as allowing problems to occur.

Maintenance problem-solving is primarily concerned with four areas: maintaining critical systems, fixing the problem quickly and faster than the last time, determining what is causing the breakdown to happen so frequently, and identifying the 20 percent of breakdowns that are consuming 80 percent of your resources.

Problems vs. Difficulties

A problem is a situation that can be characterized by a gap between your existing circumstances and where you do or do not want to be. The gap cannot be eliminated or maintained through obvious methods. Some analysis and creativity are required to define a situation as a "problem." Visualizing a problem as a gap can be a useful technique. Usually you want to overcome the gap, but sometimes you wish to maintain it. An example would be painting an object to prevent deterioration.

If you can see a solution and all it takes is good planning, then the situation confronting you should be termed a "difficulty" rather than a problem. Of course, if you are experiencing many of these difficulties, there may be a common root cause that could define a problem.

Where Maintenance Issues Originate

Issues are caused by your goals or a lack of them. You may have an overall goal of wanting your plant to run efficiently with few interruptions, but unless you translate that general goal into viable subgoals, you will experience problems. Establishing specific subgoals is essential if you wish to control the magnitude and number of the inevitable problems. Otherwise, having no goals or only general ones will magnify those problems. Often a disturbance (problem) will force you to ask, "What (unrecognized) goal do you have that is being thwarted by this situation?" Asking this question may cause you to reassess the goal.

Types of Maintenance Problems

The four common types of maintenance problems can be categorized as identification, cause/effect, means and ends.

Identification

When you don't understand a natural phenomenon, a question or a method of doing things, your natural inclination is one of curiosity. Industrial maintenance is the same way. You must identify (understand) everything in your department or plant or have someone on staff who does. When a problem occurs, you need to identify where and when it happened as well as where and when it did not. More importantly, you need to identify why you do things a certain way while always on the hunt for a better approach.

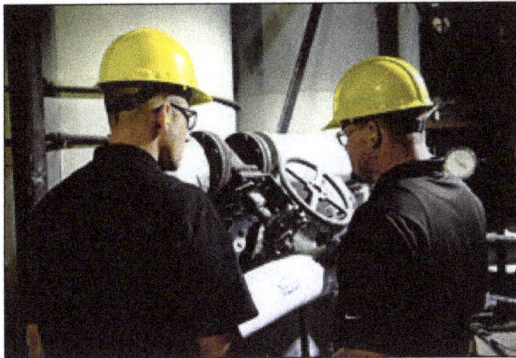

In school, you are taught the canned approach to solving problems. While this is important, it only covers problems that are recognized. What about the real-world situations? Industrial maintenance often presents situations that are so confusing that problems are camouflaged. Sorting out the mess means finding the basic problem that spawns all the other effects. This is not easy, as you may solve the wrong problem or try to alleviate symptoms caused by the basic problem. For example, you may put coolers on hot hydraulic systems instead of locating the valve or cylinder that is allowing fluid to flow back to the tank.

Identification problems become relevant not only when trying to understand a situation but also when confusion reigns and the problem is hidden by a mass of effects. The former should be attacked by curiosity and the latter by analysis. These types of problems can also appear when a manager finally asks the question, "What are we spending most of our time on and how could we minimize it?"

Cause and Effect

To properly solve cause-and-effect problems, you must first learn how to distinguish between cause and effect. Effects are things you perceive with your senses or detect through condition monitoring techniques. They accompany or precede a machine failure.

Typical effects are excessive heat, vibration and noise. A failed bearing or gear is also an effect. Simply changing the component is concentrating on the effect. While this often must be done to restore operation, forgetting about the reason for the failure is neglecting the cause. For instance, excessive heat in a hydraulic system is an effect and a predictor of problems. Concentrating on cooling the system rather than discovering the cause of the excessive heat is an invitation to problems but an all too common solution. Attack the symptom, but don't forget to unearth the root cause. Remember, symptom is a synonym for effect.

Means

Means problems are generally characterized by questions beginning with "how" such as "How can you accomplish that?" or "How can you improve that?" They leave the choice of means open-ended. With a means problem, you are trying to decide how to achieve a goal. The problem of selecting a goal or end has already been solved, so you are now focusing on how to achieve it.

Typical questions that characterize mean problems include how to reduce excessive lubricant failures, how to decrease lubricant costs while maintaining good quality, how to lessen machine downtime, how to improve safety and how to change the department mind set to prevention mode. Solving a means problem often involves finding an expert, but you should never assume the current method is the final answer. Improvement is always possible.

Ends

Problems of ends or goals can be characterized by the question, "What goal should you pursue?" your goals may be very general at first but must be translated into detailed subgoals to truly matter. Common questions to ask might include which metrics should be used to gauge progress, which 20 percent of the problems are generating 80 percent of the efforts, what are the critical parts of systems that must be constantly monitored, and how are problems categorized (critical, important and projects for correction).

Levels of Problem-solving

In addition to recognizing the four problem types, you must also be aware that problem-solving can be divided into four levels of sophistication:

- Reaction or acting on the problem when it occurs and then forgetting about it until the next time;

- Adaptation or learning to live with the problem by adjusting to the symptoms;

- Anticipation, which includes attacking root causes with preventive techniques;

- A proactive approach, which involves changing the conditions that spawned the problem in the first place.

These four levels merely describe approaches that can be used on maintenance problems. One is not better than the others but must be selected based on the severity of the problem. Of course, if a maintenance department always focuses on reaction, it might consider moving to a higher level for recurring problems.

Categories of Objectives

Your objectives will determine the problems you experience. Just as there are different levels of sophistication in problem-solving, there are different levels of objectives. These objectives are the ones you set for yourself or your department. The farther down you move on the following list, the smaller the resultant problems should be:

- Short-term Routine Objectives (Supervision): Routine objectives include maintaining things as they are, handling normal (expected) problems, reacting quickly, having lots of spares and adapting to the problem (learning to live with it).

- Medium-term Corrective Objectives (Management): Corrective objectives usually involve the elimination of accepted problems or modifying a design to solve an inherent problem.

- Long-term Improvement Objectives (Leadership): Improvement objectives might consist of requesting new equipment, changing the way things are done, concentrating on prevention and providing better training.

Most problems have an immediate phase (or crisis) and must be addressed now. However, managers who want to move to the leadership objectives will try to prevent or minimize a recurrence. While supervisors and management are concerned with doing things right, leadership concerns itself with doing the right things. Remember, setting objectives determines the problems you will encounter. Setting the right objectives will minimize those problems. In the typical plant, supervisors and management trump leadership.

Preventing Maintenance Problems

Your prevention efforts must be comprehensive and cover all areas from which problems may arise, such as personnel, maintenance practices, hardware and systems. These categories are most useful when solving cause/effect problems. However, they may also be used to keep a manager focused on all aspects of maintenance.

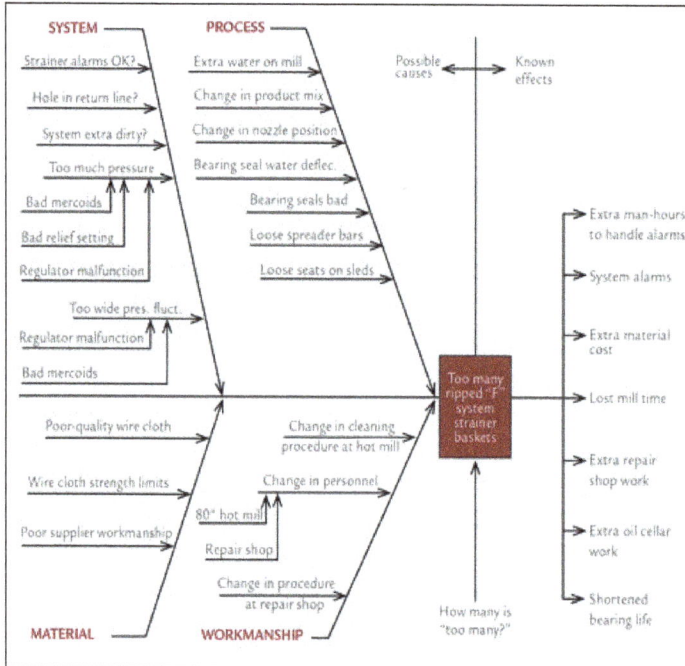

An example of an Ishikawa diagram of lube
oil system strainer basket failures.

Cause and Effect Methods

Two important techniques for establishing a problem's true cause are the Ishikawa diagram and the Kepner-Tregoe method. These techniques are especially useful with cause/effect problems that defy solution.

Kepner-tregoe Method of Problem-solving

- Compare "what should be" with "what actually is."

- The deviation is the problem.

- Identify the problem in terms of what, where (the "is"), when and extent.

- Identify what lies outside the problem in terms of what, where (the "is not"), when and extent.

- Compare the "is" with the "is not" to identify changes and distinctions.

- Find the most likely cause. The most likely cause of a deviation is one that exactly explains all the facts in the problem. If one fact can't be explained, omit that cause.

- Look for something that has changed from normal operation.

The Ishikawa diagram helps you focus on the different aspects of a problem so the listed causes will not be concentrated in one or two areas. For instance, most problems can be broken down into four areas: personnel, maintenance practices, hardware and systems. Some problems may be divisible into more than four, but with some imagination, most should yield at least these four. These categories force you to look at a situation from multiple perspectives to generate possible causes.

Some refer to these diagrams as fishbone diagrams or cause-and-effect (C-E) diagrams. They encourage you to list as many causes as possible. To do this, you must withhold judgment until the listing is complete to assure no one jumps to conclusions.

By contrast, the Kepner-Tregoe method relies on describing what the problem is, what it is not, where it occurs and where it does not. In effect, you are building a fence around the problem to keep important information inside (and under review) while keeping out extraneous information. Your main thrust is to identify what has changed. The true cause will account for all effects. If one effect could not be caused by the selected cause, that cause must be discarded.

Prevention requires maintenance management to develop a new mindset and make a conscious decision to move away from fighting fires. By understanding the four basic types of maintenance problems, the different levels of problem-solving and the three categories of objectives, you will be better prepared to achieve this new mindset.

References

- What-is-maintenance-engineering: wisegeek.com, Retrieved 26 July, 2019

- Future-maintenance-engineering, technology: manufacturingglobal.com, Retrieved 21 May, 2019

- Maintenance-organization: researchgate.net, Retrieved 8 January, 2019

- Best-management-tools-master-industrial-maintenance: mobility-work.com, Retrieved 13 May, 2019

- Common-maintenance-problems, read: machinerylubrication.com, Retrieved 16 January, 2019

3

Maintenance Planning and Scheduling

Maintenance planning and scheduling are the two major functions that are responsible for the creation of a maintenance program. Some of the maintenance scheduling techniques include six sigma maintenance, lean maintenance, computer aided maintenance, reliability centered maintenance, etc. The topics elaborated in this chapter will help in gaining a better perspective about maintenance planning and scheduling.

Planning decides what, how and time estimate for a job. Scheduling decides when and who will do the job. Planning of a job should be done before scheduling a job.

A common implementation initiative after a maintenance assessment is maintenance planning and scheduling:

- Customized or standard on-site classroom training in planning and scheduling.

- Coaching and implementation support to improve planning and scheduling on-site.

Classroom training is usually effective to increase planning and scheduling awareness, however, in order to achieve sustainable improvement training has to be followed by immediate implementation after the training or very mediocre results will be achieved.

On site and on-the-job training and implementation of a better work management process is – combined with a good condition based preventive maintenance program and an up to date accessible bill of materials – the most important process used to enable people to become more productive.

Implementation and coaching of better planning and scheduling is usually customized to client. Typical activities include:

- Define or verify existing work processes.

- Make sure all key functions such as planners, supervisors, and operations coordinators have clear roles.

- Set clear priority rules and establish meaning and criteria for existing codes.

- Improve work request usage and clarity.

- Check backlog management and clean up if needed.

- On-the-job planning enhancement for shutdowns (if applicable).

- Shutdown/Turnaround critique communication.

- On-the-job planning enhancement for day-to-day work.

- Job package creation.

- Staging areas.

- Enhance and practice planning and scheduling meetings between operations and maintenance.

- Implement Key Performance indicators (KPI's).

- Work on effective use of CMMS.

- Hand-on support for history documentation.

- Improvement of technical database (bill of materials, technical data, equipment identification).

- Improve integration with materials management.

- Contractor management and integration of work processes.

Engineer believes strongly in making sure the change process has buy-in and ownership in the organization.

It is therefore important that client's organization takes ownership as soon as possible. Engineer's role is to act as a catalyst, trainer, accelerator to coach improvements.

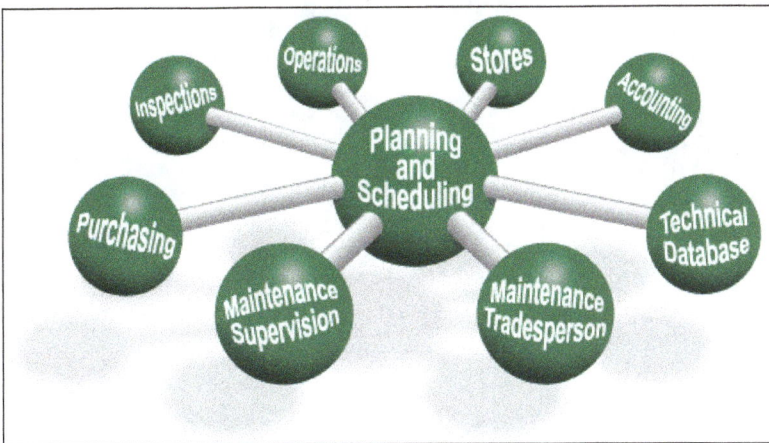

Maintenance Planning and scheduling of work orders is the hub of a well-functioning maintenance organization. In order for maintenance planning and scheduling to work many other systems need to work well. Most importantly equipment inspections through preventive maintenance, technical database such as bill of materials, work order history, and standard job plans. Maintenance spare part stores have to function well.

Effective Maintenance Planning and Scheduling will deliver a safer and more cost effective work environment for asset-intensive businesses.

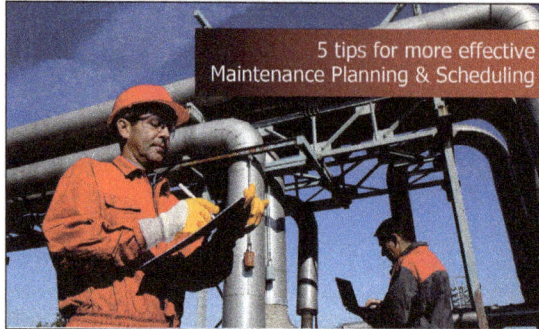

The benefits of having this function within an organisation can include:

- Safer Workplace – Planned work is inherently safer to perform than unplanned tasks.

- Labour Productivity increases up to 60%.

- Ability to collect and analyse data for failure analysis/trending.

- Accurate Budget Forecasting.

- The ability to measure workload and compliance to process (KP).

The Planning/Scheduling process can be visualised as illustrated below.

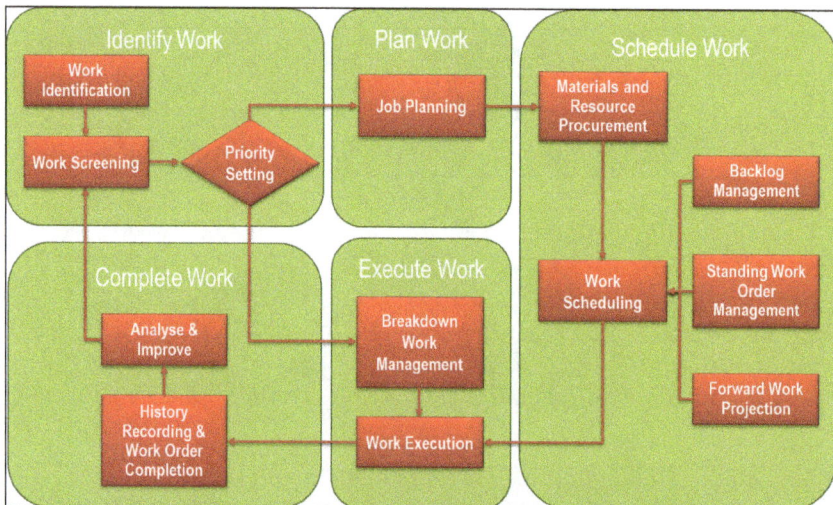

Delay Avoidance is the basic goal of Planning and Scheduling.

The prime objective of good maintenance planning and scheduling is to allow maintenance personnel to prepare and perform the required task in a safe and cost effective manner without encountering time wasting delays.

Here are five tips that can make you a more effective Planner/Scheduler.

1. Understand your Logs and what work they contain: Planners will typically utilise a large list of work that can be broken down into two categories:

- Back Log – Any work listed that has an execution date prior to today's date:

  - Typically the Backlog will contain work that was not executed before its scheduled "due by" date or have cost settlement issues that will not allow the closure of the work order. This list needs to be monitored regularly to ensure the identified work is rescheduled accordingly. These jobs will often have to be pushed into the next planning period due to a new priority status being assigned when execution is missed. Some Backlog may be unavoidable, but it should be as small as practicably possible.

- Forward Log – Any work listed that has an execution date ahead of today's date:

  - This list will be the main body of work that requires a planner's attention. This work, if it has arisen from notifications or work requests, should have been approved by someone prior to requiring planning (generally the Maintenance and Operations Supervisor). You should ensure that the job is adequately scoped enough for you to safely and cost effectively plan the job for execution, if the information is not adequate you will need to seek further information from the approver or originator of this work before you can act.

Remember the Adage: Do it Right the First Time

Some thought needs to be applied to this log and good practice is that you forward schedule these work orders so that they are scheduled to be executed as close as possible to the date that the parts (or other scarce resources) required to do the job will be available. This, obviously, means that you need to know when these will be available before you schedule the work order. In some cases, we have seen planners push the scheduled start date for all jobs that are awaiting parts to months in the future; this will cause all sorts of cost and labour resource issues within CMMS and ERP systems.

Standardise the layout of these logs wherever possible so abnormal work will stand out visually when you are looking at the work in them. Some possible ways to do this are:

- Naming Conventions, insert a prefix before the work description to segregate packages of work (e.g. "Shut Refurbish Pump 001").

- Use of Revision numbers to allocate a work week period to the tasks.

- Make use of the System and User Status codes that your CMMS/ERP system provides (e.g. the WMAT) user status code indicates that the task is awaiting materials.

These standardisation techniques will save you time and take away the pain of having to interrogate each task to understand what it means and when it needs to be done.

2. Build a culture of teamwork with your Supply Function: Parts and Materials are one of the most important aspects of successful planning and scheduling. It is a "cardinal sin" for a planner to plan and schedule a task and either not order the required parts or misread the expected delivery date. The end result is a task that cannot be performed and rework will need to be performed.

Given the importance of this a lot of organisations are ensuring that their supply/warehouse personnel attend the maintenance planning/scheduling meetings that occur each week. In some cases a Materials Coordinator role has been established to ensure there is supply/warehouse buy-in at these meetings and that missed items or delayed parts are identified prior to locking in a planned maintenance schedule. This arrangement can be extremely useful and can complement continuous improvement initiatives where a Root Cause Analysis investigation is performed whenever materials issues have caused work to be deferred. Just like planners, these Materials Coordinators have KPI measures that they will be measured against. For the business it is a win-win situation in reducing delays and ensuring that work is performed safely.

If it is not already in place, we strongly recommend fostering a great working relationship between Maintenance and Supply, as both departments will benefit from this, as will your organisation overall.

3. Measure results and process adherence: It is very difficult to visually appreciate how well a process is working unless you put some meaningful Key Performance Indicators in place. The idea of these is not to chase the "Green" compliant result but rather focus on the barriers to process adoption that are preventing the achievement of good outcomes. Often there will be circumstances that will see a KPI measure go backwards rather than the desired direction but it will alert all concerned that there is an issue that needs to be fixed in order to move forward.

Root Cause Analysis (RCA) and similar techniques can then be utilised to analyse the problem and pinpoint the issue. This may sometimes require a business case to Management for release of funds and labour to fix it. Without the accompanying measurement data it will be a very hard sell.

Once a process has been proven to work and the group is achieving the "green" result it is a great time to move onto a new measurement. There is a very real risk of complacency and you can think you are doing well but a deeper dive into the process will often show the opposite. Keep challenging the status quo of KPI measures and if they are not meaningful or relevant why are we using them?

4. Forward schedule to an agreed period: From the forward log of work there will be an expectation from every organisation that you plan ahead to an agreed period. Some very mature organisations with reliable equipment and processes will demand a longer forecast period – in this case up to 12 weeks, or even longer for major shutdown work. However, this forecast period could be as little as 2 weeks. Regardless of the length of the forecast period the aim is to have fully planned and resourced tasks that can be executed during that period. Getting all of the work in forward log fully planned and scheduled is the nirvana state as it will reduce the effort required to keep on top of this each week.

As an example: We have fully planned all work in forward log out to 8 weeks and have resourced the period to 85% of labour utilisation (to allow for breakdown work). In theory the planner will only need to keep up with new work that is generated during the week to keep the plan up to date. There will be a need to manipulate scheduled execution dates as emergent work is fitted into the forecast period. This will highlight the need for accurate priorities to be assigned to tasks. Overall, however, planner workload will diminish allowing more time to be spent on continuous improvement items.

One thing to remember is that if your forecast is for over 12 weeks of work at very high labour utilisation this can indicate that the forecast workload is out of balance with the labour available for its execution. If this occurs the planner will spend a lot of time rescheduling work orders and deferring maintenance work to the point that it is unacceptable to the business. Additional maintenance labour may be required to bring future workload back to a manageable level.

5. Analyse completed work and continuously improve: This area is where business can get the "best bang for their buck" but is often hampered by time constraints on planners as they struggle to meet schedule deadlines.

As part of the Continuous Improvement Process, maintenance planners need to analyse the work performed with the responsible Supervisor(s) and quickly make changes to any items that need attention. Performing these changes will ensure that tasks will run smoothly when executed next time.

Why do we do this? A definition of insanity is to do the same thing over and over again expecting a different result! Items that may need changing would be:

- Task Duration.

- Materials or quantity of materials.

- Work steps out of sequence.

- Labour requirements under/over resourced.

- 3rd Part requirements: scaffold, hire equipment, cranes, extra labour.

- Master data changes to strategies and plans.

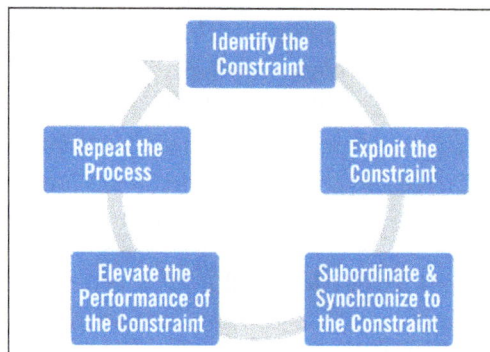

Maintenance Strategy

Often maintenance managers and maintenance engineers are asked to develop a maintenance strategy for their plant and equipment. They need to develop a document. In it you explain how you are going to use the least plant and equipment maintenance expenditure and efforts to ensure the necessary production performance from your production plant and equipment.

There is one step prior to this that needs to be mentioned. Before the development of the maintenance strategy is the development of the enterprise asset management strategy.

The enterprise asset management strategy explains the life-cycle selection and use of your plant and equipment assets to achieve the year-after-year production goals that will deliver the year-after-year business goals. Maintenance is just one of the on-going enterprise asset management strategies you will use. Maybe it's not even the most important in achieving your overall business success, but it's definitely very, very important in achieving your production success.

With the importance of maintenance to your production success firmly placed into a business context through the asset management strategy, you then need to decide how to use maintenance to maximise your production productivity. This is the role of your equipment maintenance strategy.

It is a substantial undertaking. But without it you are flying-by-the-seat-of-your-pants, everything will be guess-work. Without it vast amounts of production time and money will be wasted. You will work in an 'also-ran' operation that no one cares about. But with a maintenance strategy you have the chance of becoming one of the great companies in your industry—maybe even renowned world-wide for your incredibly successful production performance. Making your company into a world-class leader is a job worth doing well.

Typical contents of an equipment maintenance strategy document:

1. Maintenance Policy (why you do maintenance in your business, who does it, how you do it, what you expect from it).

2. Production Performance Envelope (what daily plant availability is needed to meet the production output? What is the daily average production rate that must be sustained to deliver the required output? What is the daily quality rate required to meet production plans? What equipment reliability is needed for each piece of plant to deliver the total plant availability required to meet the production plan? How much can you afford to spend on maintenance and repairs?)

   • Production Performance Required.

   • Process Reliability Analysis (reliability model your production process to identify its weaknesses and most likely performance).

3. Risk Assessment of Operational Assets (what can go wrong with your equipment, what will it cost, how often does it happen. The equation is: Risk = cost consequence [$] x no. of events in a time period [/yr] x chance of event ('chance of event' is between 1 if it will definitely happen, to 0 if it definitely will never happen). You would do this in a spreadsheet. Use the DAFT Costs as the consequences value. DAFT Cost is the Defect and Failure True Cost – the cost impact of an equipment's failure across the whole business).

   • Equipment Level (e.g. a complete pump set):

     ◦ Financial and throughput impact on production of failures of each equipment.

     ◦ Equipment Criticality (prioritise the importance of the equipment to sustaining production).

- Assembly Level, e.g. pump – coupling – motor – base frame – foundations – power supply:

  ◦ Failure Mode and Effects Analysis at part level, identify the parts in the assemblies that can fail and in which ways. Then identify the maintenance each part requires to prevent production failure.

4. Production Risk Management Plan (how maintenance is used at the parts and assembly level to reduce production risk at the equipment level):

- Precision Maintenance Standards needed to meet plant and equipment operational performance (mechanical, electrical, instrumentation, structural, civil – safety, environmental, etc).

- List Equipment on Preventive Maintenance (to adjust and replace wearing parts):

  ◦ List of equipment done as shutdown, or as opportunity based PMs, or as time/usage scheduled PMs.

  ◦ Precision standards to meet when performing PMs.

- List Equipment on Predictive Maintenance (to detect impending failure and repair/replace before failure):

  ◦ What condition monitoring will be used.

  ◦ Where will the condition monitoring be done.

  ◦ How will it be decided when it is time to maintain or replace.

  ◦ Who will do the condition monitoring (i.e. subcontract, in-house maintainer, in-house operator).

  ◦ What will be done when condition is too far deteriorated.

- List Equipment to Rebuild (to identify which equipment will be repaired):

  ◦ Criteria to pass to justify repair instead of replacement.

  ◦ How many times to rebuild before replacing with new.

  ◦ Precision standards to meet on each rebuild.

  ◦ Precision standards to meet on re-installation.

- List Equipment to Replace (to identify which equipment will never be repaired but always replaced. The DAFT Cost of a breakdown often easily justifies installing new equipment, rather than take the chance of an unplanned production stoppage):

- ◦ Precision standards new equipment must meet.

- ◦ Precision standards to meet on installation.

- • Critical Spares List (to identify which parts you must have available):

- ◦ Equipment parts to be carried on-site.

- ◦ Equipment parts to be carried by local supplier.

- ◦ Stores management standards to protect integrity of spares.

5. Records Management (to document maintenance history of equipment and parts usage in order to identify reliability improvement opportunities):

- • Which engineering, operational and maintenance documents to keep.

- • How documents are to be kept current and safe.

- • What records are to be made and kept over each equipment life.

- • What analysis of records will be required and the information to be provided from the analysis.

- • How will all the records and documents be controlled.

6. Maintenance Performance Monitoring (to ensure that maintenance is delivering the reliability, availability, quality and cost that the production plan requires):

- • KPI definitions and calculations.

- • Plant level KPIs (e.g. availability, unit cost of production, quality rate).

- • Equipment level KPIs (e.g. reliability, quality rate, production rate).

- • Personnel KPIs (e.g. hours spent developing skills, employee satisfaction).

- • Maintenance Process Performance KPIs (e.g. daily work order complete per trade type, backlog of work, percent planned work, percent scheduled achievement).

- • Maintenance Improvement KPIs (e.g. no. of procedures written to ACE 3T standard, no. of design-out projects started, no. of design-out projects completed.

- • Reliability Prediction KPIs (e.g. no. of work orders spent improving reliability, reliability improvement graphs i.e. Crow-AMSSA plots).

7. Maintenance Resources Required (there will be a need to resource the production risk management activities known as 'maintenance'):

    •   Necessary maintenance equipment and technologies.

    •   Necessary stores capacity and stores internal operating methodologies.

    •   Necessary engineering and maintenance knowledge.

    •   Necessary trade skills and competence.

    •   Necessary numbers of people by trade type/service.

    •   Location of people for most efficient operation of maintenance activities.

    •   Necessary Computerised Maintenance Management System (CMMS) capabilities.

8. Cost and Benefit Analysis (to confirm that the cost of doing maintenance will return value to the business):

    •   Annual maintenance cost verses the cost of failures prevented (the risk analysis will provide the DAFT Costs that will be incurred by the business if equipment fails).

    •   Annual maintenance cost verses the cost of lost production output if plant availability does not meet production targets (your production and equipment history can be used to determine the numbers of production slowdowns and stoppages in an 'average' year that did not need to happen).

9. Rolling Two Year Maintenance Program (indicate exactly when and what is to be done with each item of plant to deliver maximum production productivity):

    •   Work orders by type performed on each equipment item and the benefits they provide.

    •   Schedule of work orders for each equipment.

10. Rolling Two Year Maintenance Budget (you can now develop a believable budget that will deliver the risk control that your production needs. Using a rolling two years forecast lets you include the savings from improvement initiatives. Two years is a believable time period in which changes can be anticipated. A five years forecast becomes unrealistic in the later years because you cannot anticipate the impacts of a changing world.):

    •   Maintenance cost by equipment.

    •   Maintenance cost by plant.

    •   Maintenance cost by type.

- Maintenance cost per time period.

- Equipment improvement plans.

The list is reasonably comprehensive but you will need to tailor it to suit your situation and the requirements of your business and its management.

Effective Maintenance Plan

Creating a maintenance plan is generally not difficult to do. But creating a comprehensive maintenance program that is effective poses some interesting challenges. It would be difficult to appreciate the subtleties of what makes a maintenance plan effective without understanding how the plan forms part of the total maintenance environment.

Maintenance practitioners across industry use many maintenance terms to mean different things. So to level the playing field, it is necessary to explain the way in which a few of these terms have been utilized throughout this document to ensure common understanding by all who read it.

In sporting parlance, the maintenance policy defines the "rules of the game", whereas the maintenance strategy defines the "game plan" for that game or season:

- Maintenance policy – Highest-level document, typically applies to the entire site.

- Maintenance strategy – Next level down, typically reviewed and updated every 1 to 2 years.

- Maintenance program – Applies to an equipment system or work center, describes the total package of all maintenance requirements to care for that system.

- Maintenance checklist – List of maintenance tasks (preventive or predictive) typically derived through some form of analysis, generated automatically as work orders at a predetermined frequency.

- Short-term maintenance plan (sometimes called a "schedule of work") – Selection of checklists and other ad-hoc work orders grouped together to be issued to a workshop team for completion during a defined maintenance period, typically spanning one week or one shift.

Maintenance Information Loop

Figure below describes the flow of maintenance information and how the various aspects fit together.

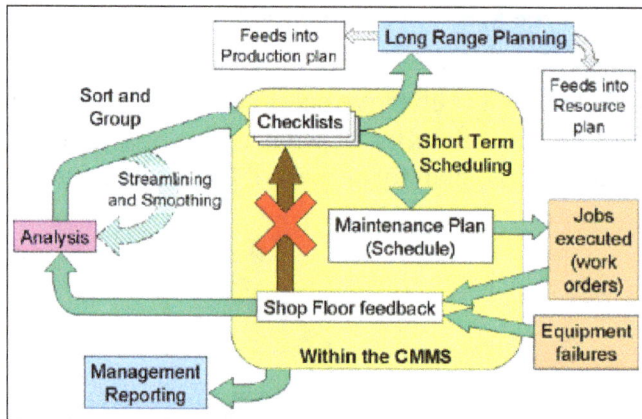

Maintenance Information Loop.

The large square block indicates the steps that take place within the computerized maintenance management system, or CMMS. It is good practice to conduct some form of analysis to identify the appropriate maintenance tasks to care for your equipment. RCM2 is probably the most celebrated methodology, but there are many variations.

The analysis will result in a list of tasks that need to be sorted and grouped into sensible chunks, which each form the content of a checklist. Sometimes it may be necessary to do some smoothing and streamlining of these groups of tasks in an iterative manner. The most obvious next step is to schedule the work orders generated by the system into a plan of work for the workshop teams.

Less common, however, is to use this checklist data to create a long-range plan of forecasted maintenance work. This maintenance plan serves two purposes:

- The results can be used to determine future labour requirements,
- They feed into the production plan.

The schedule of planned jobs is issued to the workshop and the work is completed. Feedback from these work orders, together with details of any equipment failures, is captured in the CMMS for historical reporting purposes. A logical response to this shop floor feedback is that the content of the checklists should be refined to improve the quality of the preventive maintenance, especially to prevent the recurrence of failures.

A common mistake however, is to jump straight from the work order feedback and immediately change the words on the checklists. When this happens, the integrity of the preventive maintenance programme is immediately compromised because the revised words on the checklist have no defendable scientific basis. This should be avoided wherever possible.

The far better approach to avoid this guessing game is to route all the checklist amendments through the same analysis as was used originally to create the initial checklists. This means that the integrity of the maintenance program is sustained over the long term. Implicit in this approach, however, is the need to have a robust system in which the content of the analysis can be captured and updated easily.

Finally, all the information that gets captured into the CMMS must be put to good use otherwise it is a waste of time. This is the value of management reports that can be created from maintenance information.

RCM Analysis

Without describing the complete RCM analytical process, it is instructive at this stage to point out a few details that are important to the content of such an analysis because of the way they can impact the overall maintenance plan.

Table: Information captured in the RCM-style analysis

	RCM	Additional
Identify the:	Functions Functional failures Failure modes Failure effects	Equipment hierarchy down to component level Root cause of failure
Analytical tool to select:	Failure effect category Preventive/corrective maintenance tasks (as appropriate)	
	Task frequency Crafts	Task duration Running/stopped marker

The center column is what will be found in any typical RCM-style analysis. In addition to that, there is value in constructing a hierarchy of the equipment system showing assemblies, subassemblies and individual components. This helps to keep track of which section of the system is being considered at any time, and the list of components also helps to identify the spare parts requirements for the system.

Of vital importance is the clear identification of the root cause of each failure, as this will affect the selection of a suitable maintenance task. To illustrate this point, consider for example, a seized gearbox. "Seized" is an effect. There could be several root causes of this failure mode that can be addressed in different ways through the maintenance plan. There is usually no value in aiming maintenance at the effect of a failure.

Also important from a planning perspective is to identify the time it will take to carry out each task independently. The sum total of these task times gives a good indication of how long the total work order will take. All of the above depends on the production process and the site's operating context, so these comments should be taken simply as a guideline.

The following are a few points to consider when constructing a preventive maintenance program:

Preventive maintenance tasks must:

- Aim at the failure process.
- Be specific.
- Include specifications or tolerances.

Wherever possible, aim for predictive rather than preventive tasks:

- Measure or check for conditions against a standard.
- Report the results.
- Create a follow-on task to repair or replace at the next opportunity.

"Check and replace, if necessary" tasks destroy planned times. Frequencies and estimated times for each task must be accurate and meaningful. Try wherever possible to only plan shutdown time for "non-running" tasks. Keep "running" tasks to be done during periods of normal production. Structure the maintenance program to allow for this.

Sorting and Grouping of Maintenance Plan Checklists

After analysing all the maintenance requirements for the equipment system, these individual tasks would be grouped together to create the checklists, based on common criteria for:

- Craft,
- Frequency,
- Safety/Non-safety tasks,
- Running/Non-running checks and sensible,
- Timing, etc.

Smoothing the PM Workload

In order to smooth the PM workload, a robust approach is to base the spread of PM activities on the checklists arising from the RCM-style analysis. This assumes that the analysis has been conducted thoroughly and that it is in a format that can be amended easily.

The graph in figure below illustrates how it is possible to arrange the occurrence of the PM work orders in such a way to create the smoothest possible flow of regular preventive maintenance work, while still leaving enough time to carry out those "follow-on" corrective maintenance tasks that were identified from conducting the preventive/predictive checks during the last maintenance stop.

It is important to notice that just because two checklists may have the same frequency, it is not necessary to schedule them to be done at the same time. Sometimes, of course, it does make practical sense to schedule PMs for the same day, but don't assume that this is always true. As a general rule, in an automated or continuous process production environment, the total amount of work on one checklist or work planned for one maintenance period should not exceed 80 percent of the total time available.

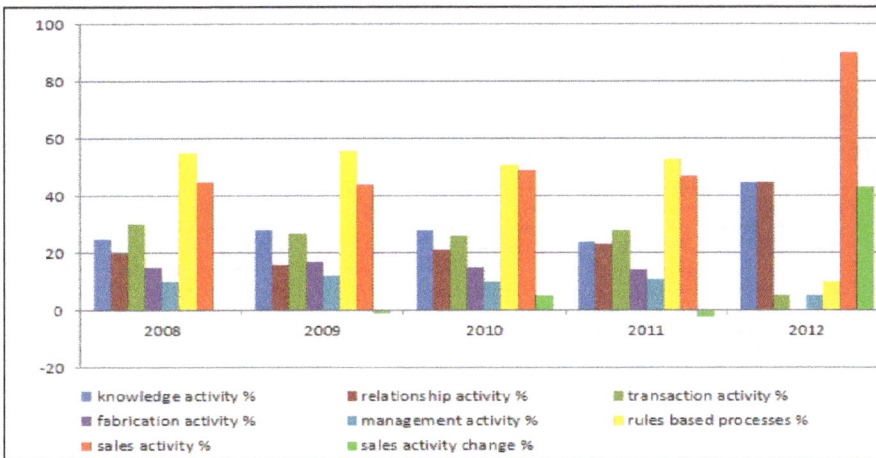

Smoothing the PM workload.

In order to achieve this smoothed workload pattern, it may be necessary to return to the timings, frequencies, groupings, start dates, etc., that were specified in the original analysis and rework some of the data. This is the iterative approach that was mentioned earlier in the description of figure.

Short-term Scheduling of Planned Maintenance Work

It is well-recognized in modern maintenance circles that there is great value in planning the maintenance workload at a macro level over a long-term horizon as well as at a detailed level over a short horizon. These two activities serve significantly different purposes.

Regular work orders are created automatically in Maximo every night from the work order templates in the PM Master table. These fresh work orders are generated typically 30 days prior to the Target Start date specified on the PM. Other work orders are also created manually by the system users, such as craftsmen and engineers.

All these work orders need to be prioritized according to the importance and urgency of the tasks, and they need to be planned into the weekly workload of the maintenance crews to ensure that a well-balanced selection of work is assigned to each crew without them becoming overloaded.

An example layout of the weekly maintenance work schedule is shown in figure below.

Maintenance Schedule

Date: _____

Prepared By: _____

Submitted To: _____

Approved By: _____

No.	Task Description	Task Duration	Due Date	Target Date	Resource Names	Person Responsible	Predecessor

Example of weekly maintenance work schedule.

Most often, a CMMS will only produce report data in text or numerical format. Because engineers like to see things in a graphical or pictorial representation, however, it may be necessary to combine the use of the CMMS with another package that has graphics capability, such as a spread sheet. The following descriptions rely on the ability of the CMMS to produce a "flat file" from a report, which can then be imported into a spreadsheet and manipulated further. If possible, it would be preferable to retain all the raw data within the CMMS and simply produce all the graphs and reports from that environment. There are two obstacles to this approach, however: Very few CMMS packages have graphical capability.

Very few CMMS packages will capture or provide the full spectrum of data that may be required to construct the desired selection of graphs. The alternative solution, therefore, is to copy the required selection of data from the CMMS to the spreadsheet environment where it can be manipulated further.

Long-range Maintenance Planning

Some sites enjoy the luxury of having regular, fixed maintenance windows built into the production plans. For example, it could be agreed that every Tuesday morning

Production Unit 1 will stop production and the equipment will be made available to the maintenance crew for six hours. During this six-hour window, the maintenance crew has the opportunity to assign as many people as required to complete all the planned maintenance activities in that work center. Thereafter, the system is handed back to the production team until the next week.

In many cases however, there is no such regular routine in place. Opportunities for the maintenance teams to conduct planned maintenance need to be negotiated and agreed with the production teams on an "as-needed" basis. Unfortunately, this is very often reduced to the maintenance department begging for access to the equipment. Furthermore, this plea is often met with the unsympathetic response from the production teams that they have to run the equipment in order to meet their targets and they therefore cannot afford to release it for maintenance.

The generation of a long-range maintenance plan that shows the number of hours of preventive maintenance work to be done in each work center over an 18- to 24-month horizon is a valuable tool. It gives the production schedulers visibility of the amount of time that is required for this preventive maintenance so that they can proactively plan to release the equipment for those periods. This makes the job of planning the maintenance activities so much simpler.

The nature of the production environment at the author's site makes it difficult to implement a regular, fixed pattern of maintenance windows as described above. For this reason a long-range maintenance plan is produced to give the production teams as much advance warning as possible of the anticipated maintenance requirements. This plan shows the forecasted maintenance hours for each operating unit, by craft type, in weekly chunks over a 24-month horizon.

Table below illustrates what the structure of a long-range maintenance plan might look like. A flat file is created from the master data table in Maximo which contains details of all the maintenance tasks and checklists with their corresponding equipment details, duration, frequencies, crafts, next due dates, etc. This information is imported into a spreadsheet, which uses a series of filters and formulae to produce the long-range plan.

Table: Example layout of long-range maintenance plan.

Week commencing		04-Jul-05 Week 1	11-Jul-05 Week 2	18-Jul-05 Week 3	25-Jul-05 Week 4
Oprating unit	Craft				
Area 1	Mech	8 M	7 M	6 M	5.5 M
	Elec	4.5 E	5 E	3 E	4 E
	Inst	1.5 I	2 I	0	3 I
Area 2	Mech	10 M	9 M	12 M	7 M
	Elec	5.5 E	3.5 E	7 E	4 E
	Inst	3 I	3 I	3 I	2.5 I

Area 3	Mech	0	1.5 M	0	1 M
	Elec	0	0.5 E	0	1 E
	Inst	0	0.5 E	0	0
Area 4	Mech	7 M	8.5 M	7 M	10 M
	Elec	4.5 E	5 E	5 E	5 E
	Inst	25. I	2 I	2 I	1.5 I

Weekly totals	Mech	25.0 M	26.0 M	25.0 M	23.5 M
	Elec	14.5 E	14.0 E	15.0 E	14.0 E
	Inst	7.0 I	7.5 I	5.0 I	7.0 I
Monthly totals	Mech	99.5 mechanical			
	Elec	57.5 electrical			
	Inst	26.5 instrumentation			

Based on this report, the production planners make the necessary allowances in the production calendars so that the equipment will be made available for maintenance. This allowance is initially made at a macro level. The exact dates and times for maintenance will be agreed in the week or two before it is due. This arrangement of the numbers can also be used to help smooth the workload across the weeks by adjusting the due dates of the maintenance tasks in the CMMS.

Long-range Labor Plan

The above explanations describe how to identify the anticipated number of maintenance hours in a production area. In order to ensure that each team on site has adequate craftsman resources available to cover all the work that will arise in their areas, a long-range workload vs. manpower forecast can be produced. This amounts to a graph that compares the hours of work to be done each month with the corresponding man-hours of labor available. A graph is constructed for each craft group within each workshop team, spanning the next 18- to 24-month horizon.

If the long-term prediction shows that the level of maintenance activity is about to increase beyond the level that can be accomplished with the existing resources, this advance warning will ensure that there will be sufficient time to recruit and train additional resources before the situation goes out of control. Similarly, a decrease in the predicted level of maintenance activity will give sufficient advance visibility of the opportunity to reassign craftsman resources to other teams or activities. This proactive approach will lead to improved manpower utilization and less panic.

Listed below are some of the categories of data that are used to construct the graphs:

1. Workload (i.e. everything that will occupy the craftspeople's time):

- Preventive maintenance hours from the CMMS.

- Breakdown allowance.

- Corrective/follow-on work/results-based tasks.

- Project work (ad-hoc hours for each forthcoming project activity independently).

- Allowances for meetings/training, etc.

2. Manpower (i.e. net man-hours available):

- Gross man-hours available in the crew.

- Allowances for leave and sickness.

- Additional allowance for overtime.

The sum of the workload hours for each month draws the workload line. The sum of the manpower hours draws the labor capacity line. Where the workload exceeds the labor capacity, the load must be smoothed, or additional resources may be required.

The preventive maintenance hours from the CMMS are obtained from the totals from the long-range maintenance plan. The allowances for breakdowns, corrective work, etc., are calculated as a rolling 12-month average of the demonstrated actual data from the CMMS. Data for other allowances may be sourced from elsewhere if not contained in the CMMS.

Manpower is basically the effective number of man-hours available for each craft in the crew.

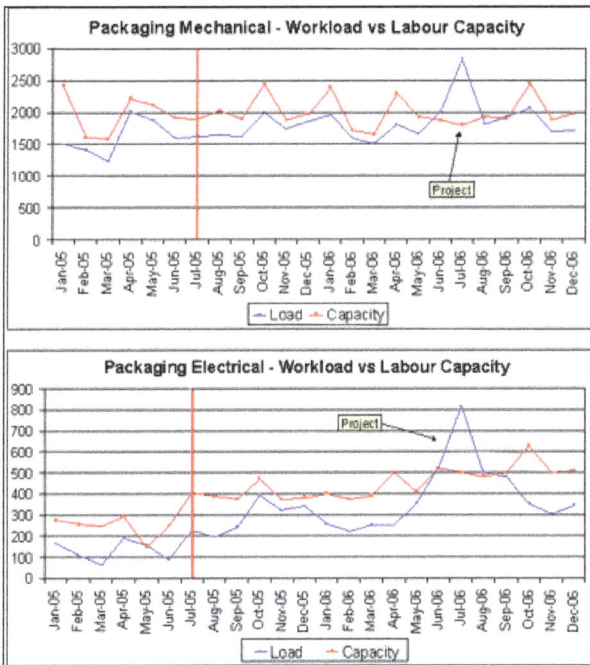

Example workload vs. labor-capacity graphs.

Where the manpower exceeds the workload, everything is in control. Where the workload exceeds the manpower, it will be necessary to reduce some of the non-essential activities at that time, or increase the people availability.

Feedback and Reporting

Feedback information returning from the shop floor, either by way of the planned work order responses, or from equipment failures will be captured in the CMMS. This information can be summarized on a report. The key recipients of these reports are the reliability engineers who look after each equipment system.

Ideally, the engineer should look at every work order that was raised in his area, but this is not always feasible, so a summary report such as this is useful. The reliability engineer must then decide on the appropriate course of action in response to each failure or observation.

Table: Example weekly failures report.

Start date 16-may -05			End Date 22-may- 05
MO25 Granulation suite 2 (and Glatt AHU)			
Equipment	16113		Collette high shear mixer
Failure	WO	Time (hrs)	
	241867	4.00	Set up wip test for collett mixer.
1		4.00	
Equipment	900955		Glatt flibeovdrier OPG2
Failure	WO	Time (hrs)	
	241895	2.00	Sieve fixed parthole leaking
	241781	1.50	Remove screen and wipe out trunking.
	241909	7.50	Broken porthole glass on sieve bow
3		11.00	
Equipment	N87		Vacumax mixer load unit
Failure	WO	Time (hrs)	
	241898	1.00	Repairs to lid seal
1		1.00	
Total: 5		16.00	

The algorithm shown in figure describes the thought process that should be going through the minds of the reliability professionals every time they review the failure work orders as shown on the summary report in table above.

It must be remembered, however, that every time the "Amend Checklists" option is selected, this amendment should be routed through the original RCM analysis to ensure the integrity of the maintenance program is not violated. Amending the checklists without running through the method and structure of the original analysis is a mistake. Regardless of the approach that has been used to record the original analysis, it is worth it in the long run to force the reliability engineers to route every amendment through the analysis and record the results for future reference.

If a spreadsheet has been identified as the most appropriate option, then it should be structured in a robust and user-friendly fashion. If it is clumsy to update, it will fall into disrepair, and the integrity of the program is lost. A database system is a far better option for this purpose, if a suitable one is available.

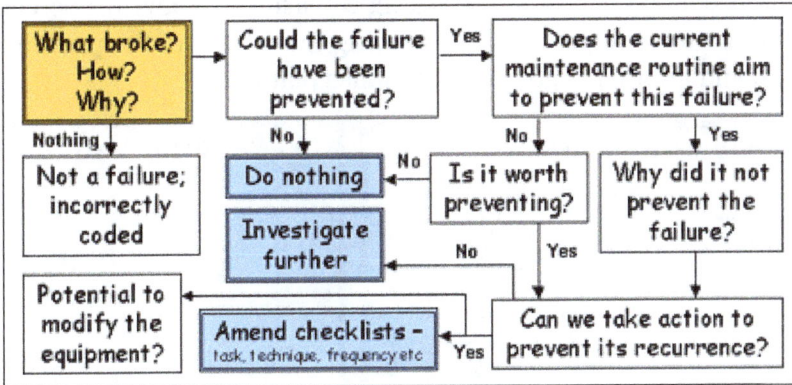

What broke algorithm.

The purpose of maintenance measures should be to monitor the health of the maintenance organisation. Where everything is in control, the metrics will reflect the success that has been achieved. Conversely, they should also be used to highlight problem areas and irregularities in order to drive the desired behaviours or areas for improvement.

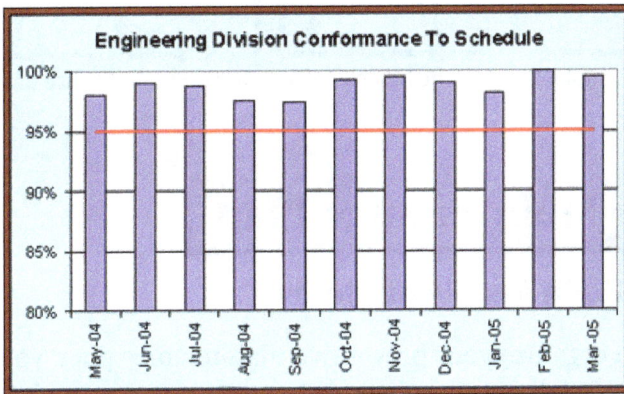

The graphs in figure illustrate some of the benefits that have been realized on the author's site as a result of having a well-functioning maintenance organisation. These graphs form just part of the regular reporting metrics by which the maintenance activities are managed. The first graph shows the conformance to the weekly planned maintenance schedule. The target is set at 95 percent and is consistently being exceeding across all of the engineering teams.

Graphs 2 and 3 show how the number of failures has been decreasing month-on-month in one particular work center over the past 12 months, and correspondingly, the mean time between failures has been increasing over the same period.

The last 2 graphs show machine availability in two of the key work centers where a full re-analysis of all the maintenance requirements was recently conducted using an adapted RCM2 approach. It is clear to see how, in both cases, the equipment availability was far out of control and from the time the improvement activity was started, the availability stabilized and is now still tracking consistently above 90 percent. This has been the result of a few things: one is improving the quality of the preventive maintenance routines, and another is good maintenance planning.

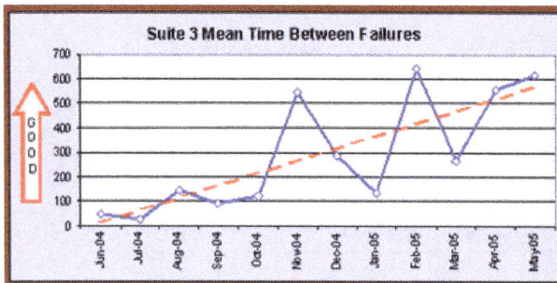

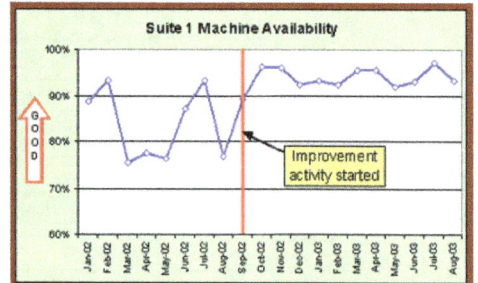

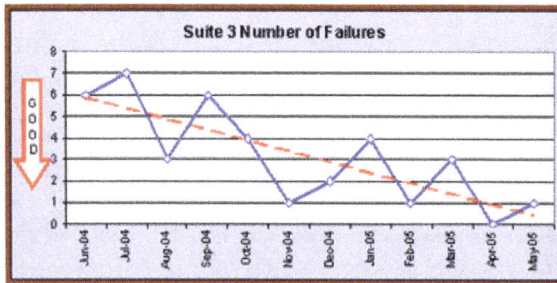

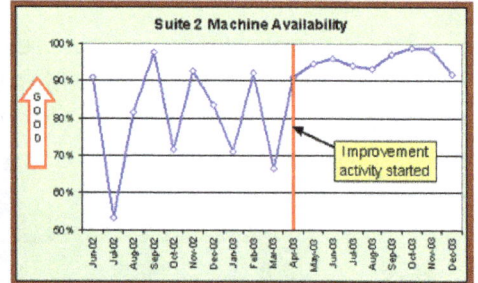

Sample graphs showing the benefits of an effective maintenance program.

Preventive Maintenance Plan

Step 1: Get the Right People on Board

Before you begin to organize your preventive maintenance plan, you need to have the right people on board with the plan. Include top management, maintenance managers, maintenance technicians, and any other staff who understands the way your system operates. This could include people from data processing, accounting, craftsmen, and members of production and production control.

You may not need input from each of these people at every step of the process, but it's important to have them on board and kept up to date so you can get important feedback as you go.

Step 2: Set Goals for your Preventive Maintenance Plan

Using your task force's input, set goals you hope to achieve using the system. Begin training your task force on the computer skills they'll need when your preventive maintenance plan goes into full effect.

Step 3: Inventory the Equipment and Assets

Go through your facility and inventory all the equipment you're considering including in your preventive maintenance plan, tagging the equipment as you go. Create a list of all the assets you have responsibility for. Record the following details as you go, and keep in mind that this process is much easier to carry out and organized with the assistance of a good preventive maintenance software program.

- Make and model of the equipment.

- Serial number.

- Basic specification and capabilities.

- Asset number, brass tag number, or unit number.

- Category (HVAC, plumbing, etc).

- The location of the equipment.

- The department who holds responsibility.

- Any high cost items of the asset.

This information will help you later track costs and help determine whether a piece of equipment needs to be replaced now.

Take note of the equipment's current condition, and rate its level of priority in relation to your overall operation.

Step 4: Make Decisions

What is the health status of each piece of equipment? You can determine this by asking these questions:

- Is it operating to manufacturers' specifications?

- Is it a high priority asset?

Once you've used these questions to analyze your equipment and have determined your highest priority asset, determine how your assets are performing and set a reasonable operational goal. Keep in mind that no system will ever run at a 100% average capacity. Compare your system's actual performance to your operational goals to determine which systems need the most attention.

It is important to remember that not every piece of equipment should be added to your preventive maintenance plan. Some equipment is just too old and worn out, and reactive maintenance may actually be a more cost-efficient method in these cases. Look at the cost of repairs or replacement, how often this maintenance is typically performed, and what level of priority the equipment has.

Then create a task list per piece of equipment. Include approximate labor time and skill level required for each task in the plan. If you're going to need parts for the task, make sure you budget for it and plan for delivery time.

A good candidate for inclusion in your preventive maintenance plan will have the following characteristics:

- The repair/replacements costs are high.

- Maintenance has to be performed routinely.

- The equipment is critical to your company's success.

You'll find that some items are better left to be replaced or repaired once they break. If so, make sure to schedule modernization of those units. If possible, plan to retire bad units. It is a good idea to actually leave bad units off the system since nothing will be done for them between inspections.

Get to Know the Owner's Manuals

Now that you have a list of candidates for your preventive maintenance program, you need to determine what is best for each piece of equipment individually. Read up on the manufacturer guidelines, as well as the warranty conditions to help you figure out the best tasks for preventive maintenance.

Decide which preventive maintenance clocks you'll be using. The clock will indicate wear on your piece of equipment. You can set the clock by number of days elapsed, run-time hours, yards or tons of product, etc. Preventive maintenance software allows for these clocks to be set easily, taking a huge task off your plate.

Schedule for Long Term Preventive Maintenance

You want to aim to get as many of your high cost/high priority pieces of equipment on a preventive maintenance schedule, but you don't have to do it all at once. Your tasks should be directed to how the unit might fail, and your goal should be to prevent as many failures as possible.

Start with one piece of equipment and add as you go. Using the first piece of equipment, create a schedule for the year, broken down into daily, weekly, monthly, quarterly, bi-annual, and annual tasks. You can use the manufacturer's guidelines to help you determine your schedule. Continue with each piece of equipment until everything has a long-term plan. Preventive maintenance software will allow you to auto-schedule preventive maintenance based on elapsed time periods or other readings, and will automatically generate an editable schedule of pending and active work orders.

Regularly review your reports from your preventive maintenance software to watch for items you need to plan for. Your long-term operating procedure will likely be revised many times throughout your first year.

Schedule for Short Term Preventive Maintenance

Now that you have planned your year, you can more easily create weekly plans for your maintenance crew. Plan preventive maintenance tasks to be performed at pre-planned equipment downtime, but give yourself some flexibility for work orders that come in from preventive maintenance inspection or reactive maintenance needs.

Prioritize the tasks, eventually creating a balance by adding or subtracting maintenance tasks and crew members. You'll enjoy a longer life on your equipment, lower maintenance costs, and shorter downtime.

Training

Don't let the careful planning and hard work go out the window by neglecting to train the people who will be working with and managing the maintenance of your equipment. Talk to each machine operator, and demonstrate correct procedures for daily maintenance and adjustments. Train them in service and repair procedures, and make sure they understand how to safely use the equipment. Use simple log forms so that machine operators will use them. Schedule a few minutes before and after each shift to inspect, lubricate, and clean up.

Importance of Planned Maintenance

Manufacturers often use the terms preventative and planned maintenance interchangeably. Routine maintenance is perhaps a more accurate term. It refers to scheduled maintenance for your equipment and facilities that can prevent equipment from prematurely breaking down. All equipment eventually wears out. The goal of planned maintenance is to keep your equipment operating as efficiently as possible, for as long as possible.

Although you can use an old-fashioned printed calendar system, service management software offers a better way to manage planned maintenance. With a sound schedule of planned maintenance activities at your manufacturing facility, you can prevent costly delays resulting from defective equipment. Let's take a look at what should be on the average planned maintenance schedule, and why it is important.

What should be on a Planned Maintenance Schedule?

No two manufacturing companies will use the same schedule because each uses a different set of equipment. Food manufacturers may have items like changing grease filters, fryer maintenance, and boiler service as part of their maintenance schedule, while an automotive manufacturer may need to run diagnostics on robotic equipment, calibrate instruments, and change air and paint filters in industrial manufacturing environments.

The most common items to include on a manufacturing company's planned maintenance schedule include:

- Filters - This includes office air filters, HVAC filters, grease and baffle hood filters, paint filtration parts, and water filters.

- Belts - Belts may need to be changed on conveyors and other equipment. Belts should be inspected frequently for signs of wear and changed before they break.

- Vehicle maintenance - If you maintain a fleet of vehicles, planned maintenance such as state inspections, oil changes, tire rotation, and cleaning can be added to service management software to ensure it is conducted on a reoccurring schedule.

- Calibration of instruments - Delicate instruments may need to be re-calibrated after a certain number of uses.

- Compressors - Compressors should be inspected to make sure there are no cracks or wear that can lead to safety hazards.

- Cleaning and changing lights - Light fixtures can get dirty and reduce the light reaching production areas. Bulbs that burn out need to be replaced to ensure that there is adequate light in the warehouse, distribution center, and manufacturing areas.

- HVAC maintenance - Air conditioning and heating systems should be routinely cleaned and checked to ensure energy efficiency and healthy air flows throughout your offices and manufacturing facilities.

- Pest control - Some industries may face unpleasant nuisances such as insects, rodents, etc. If pest control services such as termite inspection, bedbug treatments, or other treatments are required, these can also be scheduled for routine maintenance.

These are just a few areas of routine maintenance to consider. Your own business may have different needs. You can use your service management software to set up a schedule that's just right for your company.

Importance of Planned Maintenance

There are several reasons why planned maintenance is important:

1. Improves equipment efficiency: Equipment such as HVAC systems, fryers, paint sprayers, and other items benefits from regular maintenance. Clean and well-maintained systems work better. They save money by using less energy. They can do more in less time than dirty, worn out equipment.

2. Saves money: You'll save on expensive emergency repairs when your equipment is maintained on a routine schedule. Of course, you cannot prevent all breakdowns but, as you probably know from owning a personal vehicle, routine maintenance often saves on costly repairs later.

3. Boosts productivity: An unexpected equipment breakdown can seriously derail scheduled production days. Shifting order production to other equipment or having equipment stand idle while waiting for service and repair decreases productivity. Planned maintenance can reduce such issues to a minimum by addressing problems and maintenance needs before they become critical.

Planned maintenance can be a powerful ally in your quest to reduce costs and boost profits. Software such as Technisoft service manager, a Sage-endorsed solution, works with Sage ERP for a powerful, comprehensive system that helps you manage your business more efficiently.

Signs of Ineffective Maintenance Planning and Scheduling

1. You constantly reward people for going above and beyond the call of duty to put out fires. At first glance, this might not seem like a big deal, but it is if you're rewarding people solely for reactionary work and not preventive work. Of course, it's vital to recognize and reward people when they go to extra efforts to resolve a crisis – but does your organizational culture favor crisis resolution over stable, predictable, planned maintenance? Why not reward people for preventing failures?

If you find that your team is constantly doling out praise for quick thinking and hard work on reactive maintenance, that's a sign that you haven't implemented the proper maintenance planning and scheduling best practices to move your organization towards preventive – or even predictive – maintenance.

2. Maintenance and Operations are not in sync. As a result, schedules are constantly being modified". Lack of communication and repeatable prioritization processes can lead to friction between Maintenance and Operations departments. Without defined processes, accountability, and the right tools to support workers, it can be difficult to

reach consensus on how to prioritize work, leading to schedules being changed constantly as work is re-prioritized and different decision-makers vie for the work they consider most important. So, the schedule gets changed.

This silo-scenario means that few people – if anyone – sees the bigger picture and the impacts of maintenance decisions on the larger organization, because they are focused solely on their priorities and the work that affects them and their own goals.

3. Morale is suffering due to conflicting schedules: Morale is intangible, but you can feel it. Working in a reactionary mode can increases stress and may lower employee satisfaction.

Nobody wants to be in a situation where they are continually reacting to emergencies; it's stressful and demoralizing, which causes your technicians to work less efficiently. Add to that, conflicting priorities after a schedule is set and you have a minor skirmish happening between operations and maintenance on what can be done that always ends in some finger pointing.

People inherently want to do a good job and take pride in their work. The right Planning and Scheduling tools not only enable Supervisors to hold workers accountable, but also show when and why schedulers are being broken. This presents an opportunity to improve the process and get out of a reactive mode.

4. Technicians spend too much time on non-value add tasks resulting in a wrench time below 35%. "Wrench time is a measure of the efficiency of your actions." You've heard me say this before. But what does it mean?

Wrench time is how long a maintenance worker spends performing a value add task, as opposed to all the other work and time-spent not on the task, such as travel, parking, setting up, gathering materials, reviewing instructions or waiting for parts, assignments, or for an asset to be taken offline. Some of this is unavoidable, but minimizing the wasted time will make a difference. Out of an 8-hour workday, the average technician's wrench time is less than three hours per day. That's a lot of wasted time.

The average company has a wrench time hovering between 25-35%. Increasing that productive time requires efficient planning, meaning that assets are ready to be fixed, parts are in house, tools are where they need to be, etc. Effective Planning and Scheduling reduces that wasted time and increases wrench time, allowing that same technician to complete more tasks in a given day. The end result is an organization getting more value from their maintenance dollars.

5. The same machines need to be repaired frequently, because the repairs are always rushed: Maintenance workers feel like they are living Groundhog Day when they are repeatedly called to fix the same asset. When they're called out for an emergency quick fix, they often have no other option but to stick a metaphorical band-aid on it, often because they don't have the proper tools or parts, or enough time, to do a proper fix.

At first glance, this can appear to be a maintenance problem, but in actuality, it is also a symptom of inefficient Planning and Scheduling processes. Take a look at the data. Are there job plans associated with reactive work? Probably not. Was the effort to repair the failure recorded and analyzed? Were any improvements made to the process?

6. There's an emergency far too often and downtime is too high: Unplanned reactive work and unscheduled downtime is bound to happen within any organization in virtually any industry. Emergencies happen. The questions are:

- How often should they happen?

- What is reasonable and optimal for your organization?

- Are some emergencies preventable?

- Is your organization even keeping track of how frequently break-in work occurs, and why?

Whenever an asset is down for unscheduled maintenance, it is going to directly impact the company's bottom line. Effective Maintenance Planning and Scheduling contributes to keeping machines running at full capacity and reducing downtime.

Planned Maintenance Percentage (PMP) measures the efficiency of a maintenance department and is probably the best way to also determine emergency trends. PMP is the percentage of planned maintenance hours, versus overall maintenance hours. In other words, if a company spent 80 hours out of every 100 on planned maintenance, the remaining 20% or 20 hours was used on unplanned, emergency maintenance.

Generally, the most efficient maintenance departments achieve 85% of their time on scheduled maintenance. If your organization isn't achieving that – or isn't measuring and therefore can't say what your PMP is, that could be a problem.

7. You're swamped with a list of activities that should be documented for later reference, but it never gets done: In other words, maintenance staff reinvent the wheel every time they fix an asset, because there is no time to document that knowledge when similar work arises.

Documenting work management processes, roles and responsibilities, and job plans are a critical step for maximizing your department's efficiency. Understanding who does what allows people to focus on their part of the process. Having properly documented job plans take out a lot of the guesswork, for example:

- How long is this job going to take?

- What documentation needs to be in place?

- What safety concerns have been identified, including all lock-out and tag-out incidences?

- Is there a specific plan, including steps to complete the maintenance?

- What tools and parts are needed?

Does this sound familiar? "I don't have time to create job plans, because we're always dealing with reactive maintenance." Why are we always dealing with reactive maintenance? "because we don't have solid job plans in place."

8. You have no schedule compliance report card – or the one you do have is suspect: Schedule compliance is tough, because it can be so contradictory. Consider this:

- The best way to get more work done is to schedule *more* work.

- The best way to achieve schedule compliance is to schedule *less* work.

Scheduling compliance goals need to be carefully considered and designed so that they:

- Encourage the desired behavior.

- Are integrated and aligned with business processes and operating measurements.

- Are quantifiable and balanced.

- Are set at the organizational level (the level where decisions that affect the measurements are made).

But many organizations aren't even collecting or tracking the data to monitor their schedule compliance – so they have no idea how frequently break-in work is disrupting the schedule. Even worse, in some cases we don't have an accurate count of total available resource hours and the hours on our work orders are not as accurate as they could be. So, how accurate is our scorecard? It's not easy, but it is worth the effort and you will see improvements.

9. You can't keep spare parts in stock: When assets break down, there is a flurry of activity to get it back up and operational as soon as possible. This often means technicians are hitting the storeroom on a continual basis to grab spare parts, or parts must be brought in quickly, at a higher cost. It isn't cheap to keep a large inventory of spare parts, but plants that are constantly working in a reactive state need to keep these parts in house for when the machine breaks down.

Conversely, with proper Planning and Scheduling, organizations can prevent failures. When this happens, a large inventory of parts isn't necessary because purchasing has a sufficient heads-up on when to bring these parts in-house.

10. Unplanned overtime is the rule, not the exception: Overtime is costly in many senses of the word; it can affect employee morale and impact the company's fiscal health. Every organization needs overtime on occasion – but when you're in a reactive state, overtime becomes the norm. Better Planning and Scheduling means that people will be

able to get more done in the time they are allotted because maintenance will be carried out as planned, and not as needed. This should reduce the need for unplanned overtime and the budgetary blowouts that come with it. If you are scheduling overtime to handle backlog work, and the back log work orders are celebrating birthdays, there is a light at the end of the tunnel.

And just for good measure, here's one more sign that your enterprise maintenance Planning and Scheduling could use improvement:

11. Bonus: It is a challenge to create schedules, taking time away from the value-add tasks your Planners and Schedulers should be doing. We would like to say that it's not about having the hours, it's about how you spend them. If you have an hour of maintenance time, that time is going to go by – whether it's spent productively or whether it's wasted. This is where efficiency comes into play; it dictates how well your time (and therefore your money) is spent.

Poor or non-existent Maintenance Planning and Scheduling practices lead to wasted time in two different but important ways:

- The maintenance technicians are less efficient and effective because the plans and schedules weren't optimized.

- The Planners and Schedulers are less efficient and effective because the tools and processes the have are complicated and difficult to use.

In both cases, the result is that inefficiency means more time is spent (needlessly) on low value-add tasks, leaving less time to spend on high value tasks. Whether this means a maintenance worker spends more time driving to the worksite than actually fixing the asset, or the Scheduler spends more time jumping between applications and spreadsheets just to make a schedule, it comes down to the same thing: they have less time to spend on what they should be doing, tasks that add value to the organization.

Do's and Don'ts of Maintenance Planning and Scheduling

The cornerstones of effective maintenance management are maintenance planning and scheduling, which ensure that maintenance technicians are at the right place at the right time with the right tools. Effective maintenance planning and scheduling involve prioritizing and organizing work so that it is completed in the most efficient manner possible. The advantages of proper maintenance planning and scheduling include the following:

- More efficient use of labor hours.

- Reduced equipment downtime.

- Lower spare parts holdings.

- Faster execution of jobs.

- Cost savings.

- Improved workflow.

- Reduced injuries and stress.

The following are some do's and don'ts for maintenance planning and scheduling to keep in mind.

Do's of Maintenance Planning and Scheduling

- Pick the right person to be a maintenance planner: A maintenance planner is one of the most critical members of your maintenance organization. Placing the right person in this position is crucial. Planners must win the trust of maintenance technicians in order to have their work plans carried out effectively. The person you hire should be a highly skilled, qualified, and experienced maintenance technician with sound knowledge of maintenance planning principles and practices.

- Properly train the maintenance planner: Once you've selected the right person for the job, make sure that he or she is properly trained. Maintenance planners must know how to use facility maintenance management software properly. They also need to be able to extract data from the software and generate reports.

- Know the difference between planning and scheduling: Some maintenance planners don't understand the difference between planning and scheduling. Planning consists of figuring out what you're going to do, how you're going to do it, and what parts and resources you'll need to perform the work efficiently. Scheduling consists of determining when you're going to do a job. It's important for maintenance planners to plan out work prior to scheduling it.

- Ensure that maintenance planners are only planners: Don't assign multiple job responsibilities to a maintenance planner. If a maintenance planner is doing all of the things that a maintenance planner should be doing, he or she isn't going to have time for other tasks.

Don'ts of Maintenance Planning and Scheduling

- Provide poor instructions for maintenance jobs: Maintenance planners should provide detailed job instructions so that maintenance personnel can complete tasks without having to stop and search for additional information. Job instructions should include information like the estimated amount of time a job takes and the special tools and materials required. Instructions need to be simple enough for the least capable members of the maintenance crew to understand. A new maintenance technician should have the same chance of success when performing a job as a seasoned mechanic.

- Fail to provide feedback regarding the work that was done: Once a maintenance job is complete, proper feedback should be documented regarding the work that was done. If maintenance technicians merely state, "complete" or "I fixed it", it means that poor data is going into the maintenance management system and that you won't be able to identify problems and opportunities for improvement. Comprehensive feedback from maintenance technicians enables maintenance planners to learn and prepare for future jobs.

- Fail to make changes based on feedback from maintenance technicians: Maintenance planners are supposed to encourage maintenance technicians to provide feedback about preventive maintenance work orders in order to make them more effective in the future. It's important that maintenance planners take maintenance technicians' feedback into consideration and address their suggested changes, so maintenance technicians feel like their voices are being heard and continue to provide good feedback.

- Ignore key performance indicators (KPIs): Measuring the effectiveness of maintenance planning and scheduling is important. Monitor KPIs to identify opportunities for improvement and guide your decision-making process. While there is no single metric that gives you an overall view of maintenance planning and scheduling, various key performance indicators can provide insight into the performance of your maintenance organization.

Maintenance Scheduling Principles

Maintenance planning, can help your organization make the move from reactive to proactive, from wasteful to efficient, from failing to succeeding. But work planning is only the first step. Without a fully developed and implemented maintenance scheduling program, your organization can only improve so much.

Work scheduling is the process in which all required maintenance-related resources are scheduled to be used within a specific time. To ensure proper maintenance scheduling, you must account for the technician's knowledge as well as the availability of materials, tools, equipment and assets.

There are six principles that can dramatically improve your maintenance scheduling. By implementing these principles, it is possible to increase your company's productivity, regain control of your backlog, eliminate guesswork, and quickly adjust to unexpected situations and specific needs.

Let's take a deeper look at each of the six maintenance scheduling principles.

1. Plan for the lowest required skill level:

- A job plan is a documented description of the job steps. Each job plan should clearly identify the skills necessary to perform the work, and schedulers should take into consideration any special crafts, materials, tools and resources required.

- The plan should stipulate the number of technicians, work hours per skill level and the total duration of the job.

- Always assign two workers, and never estimate based on half or whole increments of a shift.

- Planning for the lowest skill level required reduces the chance that you'll send two of your best workers to a job that only needs one expert and one helper. Ultimately, you increase a technician's wrench-on time and the ability to complete other work orders.

2. Prioritize daily and weekly schedules:

- It's vital that you stick to your daily and weekly schedules; however, there will always be more work coming in and sometimes, it's unplanned.

- Define what constitutes emergency work, and document a process for how you will prioritize and handle in-progress, non-urgent and emergency work.

- If a schedule is disrupted, to reduce the impact, it's best to postpone a job that hasn't been started rather than interrupt a job that is currently in progress.

3. Schedule based on the forecasted hours available for the highest skill level:

- Each crew should have a weekly schedule that's based on the forecasted hours available for the highest skill level, which should be provided by the crew's supervisor.

- The schedule should include the job priority level and the job plan and should be based on how much work the crew can realistically finish.

- Scheduling from the forecast of highest skills available allows crews to accomplish more work by reducing coordination delays.

4. Assign work for every work hour available:

- When scheduling, identify jobs that can be easily interrupted by emergencies and urgent, reactive jobs without impacting the overall process too much.

- When 100 percent of available hours are scheduled, you can view different metrics and key performance indicators of your schedule. If you under-schedule, you're simply building inefficiency into the schedule. If you overschedule, you're increasing the likelihood of poor performance on schedule compliance.

5. Develop the daily schedule one day in advance:

- The supervisor should plan the next day's schedule based on the current jobs in progress, the weekly schedule, and any new high-priority and unplanned work.

- When scheduling the current day's work, the supervisor should match the technician's skills to their tasks.

- The supervisor should be prepared to address any emergency work that arises.

6. Measure performance with schedule compliance:

- Schedule compliance measures how well your organization followed the weekly schedule of jobs scheduled and jobs started.

- The best way to measure scheduling performance and workforce efficiency is through wrench-on time.

- When work is planned prior to assignment, you reduce the coordination delays during and between jobs.

The entire planning and scheduling program should include work identification, work planning, work scheduling, work execution, work completion and work analysis. When you have both asset reliability and maintenance reliability, you'll see an increase in both effectiveness and efficiency.

To support your team in implementing the proper scheduling principles, you'll need to make sure all your processes are documented. Additionally, everyone within your organization should have a clear understanding of the definition of a planned job.

Improve Overall Performance Metrics

When you follow the six maintenance planning principles and the above six maintenance scheduling principles, you can increase wrench-on time from 35 percent to 65 percent. That means a technician working an eight-hour day will go from completing less than three hours of actual work to completing more than five hours of actual work.

Instead of running a reactive company where you only fix something after it fails, you'll be running a maintenance center of excellence where everyone from the technicians to the supervisor shares a vision; supply, operations and engineering are integrated; system performance is constantly improving; and organizational metrics are aligned.

Maintenance Schedule Techniques

Different types of schedules are made suiting the respective job plans and different techniques are used for making and following those schedules. The first step of all scheduling is to break the job into small measurable elements, called activities and to arrange them in logical sequences considering the preceding, concurrent and succeeding activities so that a succeeding activity should follow preceding activities and concurrent activities can start together.

Arranging these activities in different fashion makes different types of schedules. They are as follows:

1. Weekly general schedule is made to provide week worth of work for each employee in an area.

2. Daily schedule is developed to provide a day's work for each maintenance employee of the area.

3. Gantt charts are used to represent the timings of tasks required to complete a project.

4. Bar charts used for technical analysis which represents the relative magnitude of the values.

5. PERT/CPM are used to find the time required for completion of the job and helps in the allocation of resources.

Modern Scientific Maintenance Methods

Reliability Centered Maintenance

Reliability centered maintenance (RCM) is defined as "a process used to determine the maintenance requirements of any physical asset in its operating context".

Basically, RCM methodology deals with some key issues not dealt with by other maintenance programs. It recognizes that all equipment in a facility is not of equal importance to either the process or facility safety. It recognizes that equipment design and operation differs and that different equipment will have a higher probability to undergo failures from different degradation mechanisms than others. It also approaches the structuring of a maintenance program recognizing that a facility does not have unlimited financial and personnel resources and that the use of both need to be prioritized and optimized. In a nutshell, RCM is a systematic approach to evaluate a facility's equipment and resources to best mate the two and result in a high degree of facility reliability and cost-effectiveness.

RCM is highly reliant on predictive maintenance but also recognizes that maintenance activities on equipment that is inexpensive and unimportant to facility reliability may best be left to a reactive maintenance approach. The following maintenance program break-downs of continually top-performing facilities would echo the RCM approach to utilize all available maintenance approaches with the predominant methodology being predictive:

- <10% Reactive.

- 25% to 35% Preventive.

- 45% to 55% Predictive.

Because RCM is so heavily weighted in utilization of predictive maintenance technologies, its program advantages and disadvantages mirror those of predictive maintenance. In addition to these advantages, RCM will allow a facility to more closely match resources to needs while improving reliability and decreasing cost.

Advantages

- Can be the most efficient maintenance program.

- Lower costs by eliminating unnecessary maintenance or overhauls.

- Minimize frequency of overhauls.

- Reduced probability of sudden equipment failures.

- Able to focus maintenance activities on critical components.

- Increased component reliability.

- Incorporates root cause analysis.

Disadvantages

- Can have significant startup cost, training, equipment, etc.

- Savings potential not readily seen by management.

Initiate Reliability Centered Maintenance

The road from a purely reactive program to a RCM program is not an easy one. The following is a list of some basic steps that will help to get moving down this path:

- Develop a master equipment list identifying the equipment in your facility.

- Prioritize the listed components based on importance to process.

- Assign components into logical groupings.

- Determine the type and number of maintenance activities required and periodicity using:

  ○ Manufacturer technical manuals.

      ◦   Machinery history.

      ◦   Root cause analysis findings, why did it fail?

      ◦   Good engineering judgment.

- Assess the size of maintenance staff.

- Identify tasks that may be performed by operations maintenance personnel.

- Analyze equipment failure modes and effects.

- Identify effective maintenance tasks or mitigation strategies.

Six Sigma Maintenance

It is the application of six sigma principles in maintenance. Six sigma is a maintenance process that focuses on reducing the variation in business production processes. By reducing variation, a business can achieve tighter control over its operational systems, increasing their cost effectiveness and encouraging productivity breakthrough.

Six sigma is a term created at Motorola to describe the goal and process used to achieve breakthrough levels of quality improvement. Sigma is the Greek symbol used by statisticians to refer to the six standard deviations. The term six sigma refers to a measure of process variation (six standard deviations) that translates into an error or defect rate of 3.4 parts per million. To achieve quality performance of six sigma level, special sets of quality improvement methodologies and statistical tools developed. These improvement methods and statistical tools are taught to a small group of workmen known as six sigma champions who are assigned full-time responsibility to define, measure, analyze, improve and control process quality. They also facilitate the improvement process by removing the organizational roadblocks encountered. Six sigma methodologies improve any existing business process by constantly reviewing and re-tuning the process. To achieve this, six sigma uses a methodology known as DMAIC (Define opportunities, Measure performance, Analyses opportunity, Improve performance, Control performance). This six sigma process is also called DMAIC process.

Six sigma relies heavily on statistical techniques to reduce failures and it incorporates the basic principles and techniques used in Business, Statistics, and Engineering. Six sigma methodologies can also be used to create a brand new business process from ground up using design for six sigma principles.

Six Sigma Maintenance Process

The steps of six sigma maintenance are same as DMAIC process. To apply six sigma in maintenance the work groups that have a good understanding of preventive maintenance techniques in addition to a strong leadership commitment. Six sigma helps in two principal inputs to the maintenance cost equation: Reduce or eliminate the need to do maintenance (reliability of equipment), and improve the effectiveness of the resources needed to accomplish maintenance. Following are the steps involved in six sigma maintenance process:

- Define: This step involves determining benchmarks, determining availability and reliability requirements, getting customer commitments and mapping the flow process.

- Measure: This step involves development of failure measurement techniques and tools, data collection process, compilation and display of data.

- Analysis: This step involves checking and verifying the data and drawing conclusions from data. It also involves determining improvement opportunities, finding root causes and map causes.

- Improve: This step involves creating model equipment and maintenance process, total maintenance plan and schedule and implementing those plans and schedule.

- Control: This step involves monitoring the improved program. Monitor improves performance and assesses effectiveness and will make necessary adjustments for the deviation if exists.

Enterprise Asset Management

Enterprise asset management is an information management system that connects all departments and disciplines within a company making them an integrated unit. EAM is also referred as computerized maintenance management system. It is the organized and systematic tracking of an organization's physical assets *i.e.*, its plant, equipment and facilities. EAM aims at best utilization of its physical assets. It ensures generation of quality data and timely flow of required data throughout the organization. EAM reduces paper work, improves the quality, quantity and timeliness of the information and provides information to technicians at the point of performance and gives workers access to job specific information at the work site.

Lean Maintenance

Lean maintenance is the application of lean principle in maintenance environments. Lean system recognizes seven forms of waste in maintenance. They are over production, waiting, transportation, process waste, inventory, waste motion and defects. In lean maintenance, these wastes are identified and efforts are made for the continuous improvement in process by eliminating the wastes. Thus, lean maintenance leads to maximize yield, productivity and profitability.

Lean maintenance is basically equipment reliability focused and reduces need for maintenance troubleshooting and repairs. Lean maintenance protects equipment and system from the routes causes of malfunctions, failures and downtime stress. From the sources of waste uptime can be improved and cost can be lowered for maintenance.

Computer Aided Maintenance

For effective discharge of the maintenance function, a well-designed information system is an essential tool. Such systems serve as effective decision support tools in the maintenance planning and execution. For optimal maintenance scheduling, large volume of data pertaining to men, money and equipment is required to be handled. This is a difficult task to be performed manually.

For a planned and advanced maintenance system use of computers is essential. Here programs are prepared to have an available inputs processed by the computer. Such a computer based system can be used as and when required for effective performance of the maintenance tasks. There are wide varieties of software package available in the market for different types of maintenance systems.

A computerized maintenance system includes the following aspects:

- Development of a database.

- Analysis of past records if available.

- Development of maintenance schedules.

- Availability of maintenance materials.

- Feedback control system.

- Project management.

Following are some computer based maintenance systems which can be implemented:

- Job card system: It is essential to prepare a job card for each component to record the maintenance work carried out or the work to be done. Job card shows the plant code, equipment code, the job code, the nature of the jobs, the start time and finishing time of the card, man-hour spent and etc. The use of computers facilitates the issue of job cards, recording of job history and control of manpower.

- Spare part life monitoring system: Under this system, information about a spare part such as its description, anticipated life and date of its installation in equipment is recorded. As and when a particular spare part is replaced during breakdown failures or scheduled maintenance, the updating of this information is done in their respective files stored in the computer. This helps to prepare the following reports:

  ○ Spares repeatability in various machines indicating the performance of such spare parts.

  ○ Comparisons of the actual life with the estimated life of the spare parts.

- Spare parts tracking system: In most of the cases maximum time is consumed in procurement of spare parts. The total time required to rectify the breakdown is summation of the time to identify the cause of the failure, time to determine the requirements of spare parts, time to procure spare parts and the time to rectify the failure. In a computerized system, the spare part tracking system is beneficial in getting required material at the earliest. A spare part file is created that contains the information about the material code, spare part identification number, the assembly or sub-assembly number and the place where the spare part is used. This helps in knowing the current position about a particular spare part and facilitates timely requirement for future demands.

Maintenance Capacity Planning

Maintenance capacity planning determines the resources needed to carry out the maintenance activities. Maintenance activities include preventive, predicative and emergency maintenance. Maintenance resources include materials, spare parts, and labor. Labor consists of all crafts and staff and it represent the most important and essential resource in maintenance. The key determinant of the maintenance capacity is the workload. The maintenance workload consists of two major components: planned work and unplanned work. Planned work consists of planned preventive and predictive maintenance, including planned overhauls and shutdowns. Unplanned work consists of emergency or breakdown (failure) maintenance. The first component is the deterministic part of the maintenance workload. The second component is the stochastic part that depends on the probabilistic failure pattern, and it is the main cause of uncertainty in maintenance forecasting and capacity planning.

The main objective in capacity planning is to provide maintenance capacity (resources) to meet random maintenance workload, in order to achieve several objectives that include maximizing system availability, safety, and utilization of limited resources. Maintenance capacity planning determines the appropriate level and workload assignment of different maintenance resources in each planning period. For each planning period, capacity-planning decisions include the number of employees, the backlog level, overtime workload, and subcontract workload.

The maintenance capacity-planning problem is a long-term planning problem. The typical planning horizon for this problem is three to five years. A major input to this planning problem is the accurate forecast of the maintenance load. This major input makes the problem stochastic in nature due to the change in operation levels and equipment age. Proper determination and allocation of the various maintenance resources to meet a stochastic varying workload is a complex and important practical problem.

The objective of maintenance capacity planning is to determine the optimum level of resources needed to meet the demand for maintenance work in each period. The demand for maintenance work is stochastic in nature. Effective maintenance capacity planning is influenced by many factors that include the knowledge about the characteristic of maintenance workload, availability of resources, and the flexibility in deploying resources. Availability of the right resources in terms of quantity and quality is critical for responding to changes in demand for maintenance. Demand management in terms of forecasting the maintenance load, scheduling resources, and monitoring performance is part of overall capacity planning.

The process of maintenance capacity planning can be briefly described as follows:

- Estimate (forecast) the total required maintenance capacity (maintenance workload) for each time period.

- Select a model to determine the required resources over time. These models include deterministic mathematical programming models, and stochastic queuing and simulation models.

- Assess the determined capacity plan using reliable performance measures.

- Adjust the capacity if needed.

The usual objective of maintenance capacity planning is to minimize the total cost of labor, subcontracting, and delay (backlogging). Other objectives include the maximization of profit, availability, reliability, or customer service.

Capacity planning techniques are generally classified into two main types: deterministic and stochastic techniques. Deterministic techniques assume that the maintenance workload and all other significant parameters are known constants. Three deterministic techniques that are presented are:

- Heuristic techniques.

- The modified transportation tableau method.

- Mathematical programming techniques, including linear and integer programming.

Stochastic capacity planning techniques assume that the maintenance workload is a random variable. Statistical distribution-fitting techniques are used to identify the probability distributions that best describe these random variables. Since uncertainty always exists, statistical techniques are more representative of real life. However, statistical models are generally more difficult to construct and solve. The three most popular stochastic models and approaches are:

- Stochastic programming.

- Queuing models.

- Stochastic simulation.

After determining the capacity plan using any given approach, it should be assessed regularly using reliable performance measures. The following measures are suggested to assess the adequacy of the maintenance capacity.

- Utilization of resources (labor utilization).

- Response time.

- Mean time to repair.

- Maintenance backlog.

- Inventory turnaround.

If the values of theses performance measures are outside the desired band, then an action to adjust the capacity plan is necessary.

Forcasting Maintenance Load

Prior to capacity planning or designing a new maintenance organization, an essential activity is to estimate the expected maintenance load. The load consists of the following components:

- Failure and emergency maintenance workload. This can be forecasted using actual historical workloads and the appropriate techniques of forecasting that include moving average, regression, and exponential smoothing.

- Preventive maintenance and predicative maintenance workload. This mostly planned work will be obtained from planned maintenance program. This includes routine inspections and lubrications.

- Deferred corrective maintenance. This can be forecasted based on historical records and future plans.

- A forecast for overhaul of removed items and fabrication. This can be estimated from historical records coupled with future plans for improvements.

- Shutdown, turnarounds, and design modifications. This can be obtained from actual historical records and the future maintenance schedule.

The forecasting of the maintenance load for a new plant is more difficult and must rely on similar plants 'experiences, benchmarking, management experience, and manufactures' information.

After the maintenance load is forecasted, the forecast should be evaluated using standard error measures such as the Mean Absolute Percent Error (MAPE). Based on this evaluation, the forecast could be revised if necessary.

Methods and Approaches for Capacity Planning

The approaches and models for capacity planning are presented. The focus is on deterministic models because they are more practical and easy to implement in real life. A brief outline of stochastic models is also provided.

1. Deterministic Approaches: The deterministic techniques include the transportation method, linear programming ad integer programming models. In production planning, products from the current period can be kept in inventory to satisfy demands in future periods. However, this is not the case in maintenance planning. In maintenance, unfinished work is backlogged and performed in future periods at a higher cost. It is possible to divide the maintenance load into several categories by skill or priority and to perform capacity planning separately for each category. For example, table shows a tableau for a three-period maintenance plan for one kind of maintenance work, which is the mechanical workload.

Table: Transportation tableau with 3 periods and 3 resources.

Execution Periods	Resource used	Demand Periods			Capacity
		1	2	3	
1	Regular Time	CR	∞	∞	R_1
	Overtime	CO	∞	∞	O_1
	Subcontract	Cs	∞	∞	S_1
2	Regular Time	$Cr + \pi$	cR	∞	R_2
	Overtime	$Co + \pi$	cO	∞	O_2
	Subcontract	$CS + \pi$	CS	∞	S_2
3	Regular Time	$CR + 2\pi$	$Cr + \pi$	CR	R_3
	Overtime	$CO + 2\pi$	$CO + \pi$	CO	O_3
	Subcontract	$Cs + 2\pi$	$CS + \pi$	Cs	S_3
Maintenance Workload		M_1	M_2	M_3	

The notation in the table is defined below:

C_r = cost per hour of in-house mechanical maintenance labor in regular time,

C_o = cost per hour of in-house mechanical maintenance labor in overtime,

C_s = cost per hour of subcontracted mechanical maintenance labor,

π = cost of backordering (doing work late) by one time unit,

R_t = capacity of in-house regular time in period t,

O_t = capacity of in-house overtime in period t,

S_t = capacity of subcontracting in period t,

M_t = forecasted mechanical maintenance load in period t.

If a work is backlogged from period (t) and performed at period ($t + i$) with regular in-house labor, it will have a higher cost of $Cr + i\pi$ per hour. Table shows the data needed for a three-period planning horizon. The table shows the costs, capacities and maintenance load. The symbol ∞ in the cost cell means that a work cannot be done in this period. For example, if work came in period ($t + 1$), then it cannot be performed in period t.

A simple least-cost heuristic method can be used to compute the allocation of the maintenance load to different sources of labor supply. The least-cost method provide a good solution, however if the optimal solution is needed the stepping stone method can be employed.

In general, the transportation tableau method requires the following data:

- Cost of maintenance for each source per man-hour for each period.

- Cost of advancing (early maintenance) per man-hour per unit time.

- Cost of backordering (late maintenance) per man-hour per unit time.

- Maintenance demand (required workload) in each period.

Linear programming (LP) is another deterministic model for capacity planning. LP is a mathematical model that minimizes or maximizes a linear function subject to linear constraints. The linear programming model determines the optimal values of decision variables to achieve a given objective such as minimizing cost or maximizing profit. The capacity-planning problem can be formulated as a linear programming problem. This type of model can be solved by Microsoft Excel, Linear Interactive and discrete optimizer (LINDO), International Mathematical Software Library (IMSL), the Optimization Software Library (OSL) and the well-known CPLEX package. The LP solution provides the optimal values and sensitivity analysis. If linearity assumptions are not satisfied, nonlinear programming has can be used for the optimal planning of maintenance resources.

If the crafts and skill levels are required to be specified in terms of employees to be hired in a maintenance department, we either round up the linear programming solution to the nearest number of employees or reformulate the problem as a mixed integer programming model. This formulation requires the following notation:

- F_{kt} = Forecasted maintenance load of grade k in period t.

- B_{kt} = Backlog of grade k maintenance work in period t.

- U_{Bk} = Upper limit for a healthy backlog for grade k work.

- L_{Bk} = Lower limit for a healthy backlog for grade k work.

- U_{ijt} = Upper limit on the availability of skill I maintenance employee from source j in period t.

- P_{ijk} = Productivity of a maintenance employee of skill i.

- r_{kt} = Cost of backlogging one man-hour of grade k work in period t.

- n_{ijkt} = Number of employees of skill i, from source j, assigned to perform maintenance work of grade k in period t.

- NH = number of hours worked by a regular in-house employee per period.

- S_j = 1 for j = 1: full-time in-house employees; 0 < Sj < 1 for j = 2: part-time or overtime employees.

- U_{ijt} = Upper bound on the availability of employee of skill i from source j in period t.

- C_{ijt} = Pay of an employee with skill i from source j in period t.

Assuming part-time and overtime cannot exceed 25% of full-time maintenance man-hours, the integer programming model for determining the optimum number of workers of different types and skills is stated as follows:

$$\text{Min} \sum_i \sum_j \sum_k \sum_t C_{ijt} \, n_{ijkt} + \sum_k \sum_t r_{kt} \, B_{kt}$$

Subject to:

$$\sum \sum (NH \times S_j) P_{ijk} n_{ijkt} + B_{kt} = F_{k,t} + B_{k,t-1}$$

$$n_{i2kt} - 0.25 \, n_{i1kt} \leq 0$$

$$n_{ijkt} \leq U_{ijt}$$

$$LB_R \leq B_{kt} \leq UB_k$$

$$n_{ijkt}, B_{kt} \geq 0 \text{ and integer.}$$

The software packages mentioned earlier have versions that can solve mixed integer programming models. The mixed integer programming model presented above is only a general-purpose example. Different components of this model could be added, deleted, or modified in order to tailor it to a specific maintenance capacity planning applications. The data needed to use or apply such types of models are:

- Cost of advancing (backlogging) each maintenance hour by one period, i.e., cost of early (late) maintenance.

- Cost of regular time (overtime) maintenance per hour per work type for in-house maintenance.

- Cost of subcontract maintenance per hour per work type.

- Cost of hiring (firing) one worker.

- Maximum number of hours available for regular in-house maintenance, in each period.

- Maximum overtime hours available in each period.

- Maximum subcontractors hours available in each period.

- Demand (forecast) of each type of maintenance work in each period.

- Acceptable backlog levels for each work type per period.

2. Stochastic Approaches: Stochastic approaches that may be used for maintenance capacity planning include:

- Stochastic programming.

- Queuing models.

- Stochastic simulation.

The driver behind these models is the stochastic nature of the demand for maintenance work. To use them effectively, these models require probability distribution identification and large amounts of data. Experience has shown limited use and application of stochastic models in maintenance capacity planning.

Maintenance Performance Metrics

The maintenance function is inherent to production. Even so, understanding and quantifying its activities can be problematic.

The scope of maintenance in a manufacturing environment is illustrated in its numerous definitions. The British Standards Institute defines maintenance as a combination of all technical and associated administrative activities required to keep equipment, installations and other physical assets in the desired operating condition or to restore them to this condition. Meanwhile, the Maintenance Engineering Society of Australia (MESA) indicates that maintenance is about achieving the required asset capabilities within an economic or business context. Maintenance also includes engineering decisions and associated actions that are necessary for the optimization of specified equipment capability, where capability is the ability to perform a specified function within a range of performance levels that may relate to capacity, rate, quality, safety and responsiveness. Similarly, Kelly states that the objective of maintenance is to achieve the agreed output level and operating pattern at a minimum resource cost within the constraints of system condition and safety. The desired production output is achieved through high availability, which is influenced by equipment reliability, maintainability and maintenance supportability. Finally, maintenance is also partly responsible for technical systems' safety and for ensuring that the plant remains in good condition.

One can say that maintenance objectives as the following; ensuring the plant functions (availability, reliability, product quality etc.); ensuring the plant reaches its design life; ensuring plant and environmental safety; ensuring cost effectiveness in maintenance and the efficient use of resources (energy and raw materials). With respect to production equipment, proper maintenance will set system functioning as its prime objective. In short, maintenance must ensure the required reliability, availability, efficiency and capability of the whole production system. It will ensure system life by keeping the equipment in good condition. In this case, cost has to be optimized to meet the desired plant condition. Plant safety is also very important, as failures can have catastrophic consequences. Here, the cost of maintenance has to be minimized while keeping the risks within strict limits and by meeting the statutory requirements.

For a long time, maintenance was carried out by the workers themselves, with no defined parameters. Equipment maintenance was more loosely organized, and there was no haste for the machinery or tools to be operational again. Given current concerns about money and safety, this is beginning to change. The focus is now to keep equipment operational or returning it to production as quickly as possible. The challenges are the following:

- First, there is a need for higher plant availability in a global economy. Global markets suffer from expansions, purchase of industrial buildings, production equipment, acquisitions of companies in the same sector, regardless of the country. Global competition means that companies want their productive capacities to remain at a maximum. Therefore, organizations are beginning to worry about keeping track of the parameters that may affect the availability of their plants and machinery.

- Second, the bottom line is chrematistic, i.e. related to money-making. When organizations begin to optimize their production costs, they start to question their maintenance costs. This function, in recent years, has grown in assets, personnel, etc., and now consumes a significant percentage of the overall organization budget. Thus, when establishing policies to streamline costs, the maintenance budget is a crucial part of the puzzle. At the same time, however, the organization's maintenance must meet availability and quality parameters. A constant concern, then, is maximizing availability at the lowest cost. Not surprisingly, methodologies and technologies to determine the best way to achieve this balance are increasingly popular, as noted by Al-Najjar.

Need to Measure Maintenance Performance

Today, organizations are under pressure to continuously enhance their capabilities to create value for their customers and improve the cost effectiveness of their operations. In this regard, the maintenance of large-investment equipment, which was once thought to be a necessary evil, is now considered key to improving the cost effectiveness of an operation, creating additional value by delivering better and more innovative services to customers.

With the change in the strategic thinking of organizations, the increased amount of outsourcing and the separation of OEMs and asset owners, it is becoming crucial to measure, control and improve the assets' maintenance performance. As technology has advanced, various maintenance strategies have evolved, including condition based maintenance, predictive maintenance, remote-maintenance, preventive maintenance, e-maintenance etc. The main challenges faced by organizations today are choosing the most efficient and effective strategies to enhance and continually improve operational capabilities, to reduce maintenance costs and to achieve competitiveness in the industry. Therefore, in addition to formulating maintenance policies and strategies for asset maintenance, it is important to evaluate their efficiency and effectiveness.

Maintenance Performance Measurement (MPM) is defined as the multidisciplinary process of measuring and justifying the value created by maintenance investment, and taking care of the organization's stockholders' requirements viewed strategically from the overall business perspective. MPM allows companies to understand the value created by maintenance, to re-evaluate and revise their maintenance policies and techniques, justify investment in new trends and techniques, revise resource allocations, and to understand the effects of maintenance on other functions and stakeholders as well as on health and safety etc.

Unfortunately, these maintenance metrics have been often misinterpreted and they are often incorrectly used by businesses. The metrics should not be used to show workers that they are not doing their job. Nor should they be used to satisfy the organization's ego, i.e. to show that the company is working excellently. Performance measurements, when used properly, should highlight opportunities for improvement, detect problems, and help find solutions.

In their overview of the state of maintenance, its current problems and the need for adequate metrics for its quantification, Mata and Aller note that maintenance is seen in industry as a necessary evil, an expense or loss, which the organization must incur to keep its production process operative. Because of this, the priorities of a company do not typically focus on maintaining assets, but on the production that they represent. However, the use of objective indicators to evaluate these processes can help to correct deficiencies and increase the production of an industrial plant. Many indicators relate the costs of maintaining to production or sales; others make it possible to determine whether availability is adequate or what factors should be modified to achieve its increase.

This historical view of maintenance, mixed with traditional issues of performance measurement, creates problems in the development and implementation of a comprehensive package of maintenance performance management. For example, the human factor must be included in the selection of the measuring metric, its implementation and the use of the resulting measurement:

1. Too much data and too little information: Data acquisition has become relatively simple and cheap through the introduction of modern and powerful hardware systems and software. That being said, data overload is now a problem, and sophisticated data mining algorithms are required to get useful information as argue. In instances when data are more difficult to collect, one needs to decide if their value to the company and specifically to a certain hierarchical level is worth the effort and cost. This is accomplished by establishing what is important for different levels, i.e. analyzing objectives tailored to each organizational level which emanate from the corporate levels. Once user needs are fully understood, it is possible to determine the maintenance strategy, organization, resources and systems.

2. The number of performance indicators, ownership of the data and the aspects to be covered: The number of indicators used for each figure or department should be limited by identifying key features or key factors. Scorecards with large numbers of indicators that do not define the users or responsible personnel actually hinder the work for which they are developed.

    To control the scorecard, it is important to approach the issue of data ownership and the need for collaboration with the rest of an organization. Often, the maintenance department is overwhelmed in its duties, so data cannot be collected. Further, there may be a lack of historical data, making it impossible to create certain indicators. In a multifunctional organization, it is likely that other departments are collecting some data critical to the generation of these parameters and can share them. For example, it may be relatively simple for the production department to collect data on availability or reliability. Occupational Safety and Health people are ideal for determining the rates of accidents.

3. Objectives and measures: At times, departments within the same company have conflicting interests in the maintenance of their equipment. But the purpose of the objectives is to ensure that departmental efforts are aligned with business

needs. Tangible goals should be tailored to the user and be free of ambiguity. Problems can be created when management fails to set goals at the highest level or fails to ensure that these objectives are correctly translated into subgoals at lower levels. However, ambiguities disappear, when management ensures that its objectives are translated into objectives at lower levels. Objectives should be transmitted in a downward cascade, including all company departments; the measure indicated by the selected sensors will indicate the appropriate steps to take to ensure that everyone is going in the same direction.

4. Time lag between action and monitoring results: Sometimes there is a delay between policy change and the appearance of clear and apparent results associated with that change. A second delay may occur between the appearance of results and the time that the measurement is taken. Each problem must be set against each objective, taking into account that technical levels can expect faster changes in their indicators than corporate levels, whose KPIs are slower to show visible results. Once a measure has been identified for a goal and level, and this is implemented, the method of data collection and the frequency must be tailored to the factors involved: physical, human, financial, organizational factors etc.

5. The cost and the reasons for data collecting: The success of any measurement system is based on the method used for data collection. Poor or incorrect data entered into a reporting system gives little value. Human factors involved in the collection of data are more reliable, as these data are more closely related to indicators of ownership and responsibility. Technicians and operators will collect data only if they believe it is worthwhile, and the results are made available for consultation and use.

If there is a risk that the indicators derived from the reported data are used against people, it is almost certain that they will not be collected in an appropriate way. Also if time passes and the data have not been used for anything, if they have been forgotten and feedback has not been transmitted, people will inevitably see the whole thing as a waste of time. In other words, if people understand the purpose and see the results, they will be motivated to collect data. Massive data collection can generate unknown indicators for the collectors; thus, they may distrust the data and fear their effects.

These issues reinforce the idea that the measurements should combine the internal functioning of maintenance with its interaction with external actors, particularly clients. At the same time, they must honor the objectives of management, as it is management who will propose improvements after reading the indicators.

Measurement: Sensors and Placements

Measurement is the act of assigning numbers to properties or characteristics. The measurement objective is to quantify a situation or to understand the effects of things that are observed TRADE. Measuring performance is essential in any business. Continuous improvement is the process of not accepting the status quo, as Wireman notes. Levitt agrees with Wireman and maintains that a prerequisite for the maintenance function is continuous improvement.

An increasing number of studies attempt to establish the relationship between maintenance performance and the reliability of a productive or operative system. For and Atkinson et al. measurement objectives are planning, selection, monitoring and diagnosis. Mitchell argues that measurement figures are needed to estimate the scope for competition, prioritize resources and determine the progress and effectiveness of improvement initiatives. Arts et al. see performance measurements (PM) as ways to control maintenance to reduce costs, increase productivity, ensure process safety and meet environmental standards. PM provides a base for improvement, since without measurement there can be no certainty of improvement. PM is a powerful methodology which allows engineers and managers to plan, monitor and control their operation/ business. In brief, the purpose of measuring maintenance performance is to help determine future action and to improve performance based on past data. If an organization does not select the appropriate metrics to measure performance, results could be misleading.

In TRADE, a performance measure is a number and a unit of measurement. The number gives the magnitude and the unit gives a meaning. Implementing measures may also be represented by multiple units expressed as ratios of two or more fundamental units to yield a new unit, TRADE. Some applications develop an indicator of performance measurement. An indicator, therefore, is a combination of a set of performance measurements. To streamline performance indicators, Key Performance Indicators (KPIs) are created; these could consist of several indicators and metrics. To determine performance level, the strengths and weaknesses of a strategy must be considered; accordingly, the selected KPIs must reflect this need.

An important aspect of MPM is formulating maintenance performance indicators, linking maintenance strategies with overall organizational strategy. The end user wants the fewest possible indicators to monitor the entire system, no matter how complex it may be. A review of the literature reveals that many attempts have been made to use maintenance performance measures as a means to develop an effective and efficient MPM system. The major issue in measuring maintenance performance is the formulation and selection of Maintenance Performance Indicators (MPIs) that reflect a company's organizational strategy and give maintenance management quantitative information on the performance of the maintenance strategy.

Hernandez proposes a battery of indicators from system reliability and functional safety. He defines an indicator or index as a numerical parameter that provides information about a critical factor identified within an organization, for example, processes, people, or expectations of cost, quality and deadlines. Indices should be few, easy to understand

and measurable; it should be fast and easy to learn how things are going and why. In addition, they must identify the key factors of maintenance; establish records of data allowing periodic calculation to set standard values for these indicators, mark targets based on those standards, make appropriate decisions and take appropriate actions. Hernandez places special emphasis on ranking these indicators; this is especially relevant when there is a large set of indicators.

Many authors agree that the first step is to develop maintenance performance indicators, i.e. numerical parameters on critical factors associated with measurable physical characteristics must be identified. identify seven basic characteristics that can be used to measure performance: quantity, price, speed, accuracy, function, service, and aesthetics.

Type of Indicators: Leading versus Lagging and Hard versus Soft

PIs are used to measure the performance of any system or process. A PI is a product of several measures (metrics). When used to measure maintenance performance in an area or activity, it is called a maintenance performance indicator (MPI). PIs are used to find ways to reduce down time, costs and waste, operate more efficiently, and increase the operation's capacity. A PI compares actual conditions with a specific set of reference conditions (requirements), measuring the distance between the current situation and the desired situation (target), the so-called 'distance to target' assessment. The list of PIs is long, and each organization's selection of performance indicators will reflect its corporate strategy's objectives and requirements.

PIs can be broadly classified as leading or lagging indicators. A leading indicator warns the user about the non-achievement of objectives before there is a problem. It is one of a statistical series that fairly reliably turns up or down before the general economy does. A leading indicator thus works as a performance driver and alerts the head of the specific organizational unit to ascertain the present status in comparison to the reference one. Soft or perceptual measures like stakeholder satisfaction and employee commitments are often leading indicators in the sense that they are highly predictive of financial performance. When such measures are tracked today, it leads to less worry about tomorrow's budgets.

A lagging indicator normally changes direction after the economy does. Lagging indicators are useless for prediction; the value of construction completed, for example, is outdated, as it would indicate the condition after the performance has taken place. The maintenance cost per unit or return on investment calculation could be an example of a lagging indicator.

The establishment of a link between the lagging and the leading indicator makes it possible to control the process. Furthermore, indicators should be chosen to accord with the chosen maintenance strategy.

The complexity of some measures is an obstacle to their implementation which decreases the likelihood of their use. In maintenance, many processes can be measured directly. Time or costs are quantities whose measurement is relatively easy. Others, such as the adequacy of repair shops and the size and type of the maintenance teams, are particularly sensitive and can only be measured with more complicated and subjective methods. This difference suggests that the indicators fall into two broad groups, 'hard' and 'soft'. 'Hard' indicators include those measurable through the extraction and exploitation of simple databases, like CMMS (computer maintenance management system), ERP (Enterprise Resource Planning) databases, presence, purchase orders, energy consumption by area etc. Arts, Knapp and Mann explain the development of a MPM system using the CMMS. The operational view of the maintenance function requires certain indices for performance measurement; it does not require the tactical and strategic aspects of maintenance performance. In this case, the data collection and calculation of the indicators are fast, and measurement does not interfere in the daily work of the maintenance team. Here, a common database can be an important instrument for maintenance management decision-making.

While many 'soft' indicators are interesting, their measurement can be rendered problematic by the absence of sources, their hard objectivity or their lack of reliability. Apart from staff and workshop size, this group includes all measures relating to elements with a strong human component, such as the impact of a training activity on the quality of repairs, or the time required for diagnosis and improvement, usually not quantified in records.

Thus, the choice of measures and the indicators derived from them will be conditioned by the accessibility and reliability of the sources, with special emphasis on the soft indicators that are affected by human factors.

The people who operate the equipment are a valuable source of information. The human element is indispensable in the measurement of maintenance due to its influence on repairs. However, to assess the overall status of a maintenance system and to correct critical points, more objective tools are needed. To this end, mathematical models and some indicators can be used to assess the probability that a team is performing inspection, maintenance or repair, and determine the average time for equipment to fail after a maintenance intervention.

In other words, two actors are involved in the MPM: people and mathematical models. People provide information on their links to the company, morale, training, skills, and confidence and so on; models provide information on effectiveness and efficiency related to cost or time. Combining the two leads to the attainment of the three objectives of excellence, noted by; efficiency, effectiveness and staff involvement.

Different categories of maintenance performance measures/indicators are identified in the literature. The total productive maintenance (TPM) concept, launched in the 1980s, provides a quantitative metric called Overall Equipment Effectiveness (OEE) for measuring productivity of manufacturing equipment. It identifies and measures losses in important aspects of manufacturing, namely, availability, performance/speed and quality. This supports the improvement of equipment effectiveness and hence its productivity. The OEE concept has become increasingly popular and is widely used as a quantitative tool to measure equipment performance in industries. Arts and Mann use the time horizon to classify maintenance control and performance indicators into three levels: strategic, tactical and operational. Indicators proposed for operational control include: planned hours over hours worked, work orders (WO) executed over WO scheduled, and preventive maintenance (PM) hours over total maintenance hours.

Parida proposes a multi-criteria hierarchical framework for MPM that consists of multi-criteria indicators for each level of management, i.e. strategic, tactical and operational. These multi-criteria indicators are categorized as equipment/process related (e.g. capacity utilization, OEE, availability etc.), cost related (e.g. maintenance cost per unit production cost), maintenance task related (e.g. ratio of planned and total maintenance tasks), customer and employee satisfaction, and health, safety and environment (HSE). Indicators are proposed for each level of management in each category.

Campbell classifies the commonly used measures of maintenance performance into three categories based on their focus. These are measures of equipment performance (e.g. availability, reliability, etc.), measures of cost performance (e.g. maintenance, labor and material cost) and measures of process performance (e.g. ratio of planned and unplanned work, schedule compliance). Coetzee outlines four categories of maintenance performance measures with detailed indicators for each category. The first category is maintenance results, measured by availability, mean time to failure (MTTF), breakdown frequency, mean time to repair (MTTR) and production rate. The second is maintenance productivity, measured by manpower utilization, manpower efficiency and maintenance cost component over total production cost. The third is maintenance operational purposefulness, measured by scheduling intensity (scheduled tasks time over clocked time), breakdown intensity, (time spent on breakdown over clocked time), breakdown severity, work order turnover, schedule compliance, and task backlog. The fourth is maintenance cost justification, measured by maintenance cost intensity (maintenance cost per unit production), stock turnover and maintenance cost over replacement value.

Ivara Corporation has developed a framework for defining KPIs based on their physical asset management requirements and the asset reliability process. They propose twenty-six key maintenance performance indicators and classify them in two broad categories, leading and lagging indicators. Leading indicators monitor the tasks that when performed will 'lead' to results (e.g. if the planning took place or if the scheduled work was completed on time), while lagging indicators monitor the results or outcomes that have been achieved (e.g. the number of equipment failures and down time). Leading indicators are classified as work identification (e.g. percentage of proactive work done), work planning (e.g. percentage of planned work), work scheduling and work execution (e.g. schedule compliance). Lagging indicators are classified as equipment performance (number of functional failures, safety and environmental incidents, and maintenance related downtime) and cost related measures (e.g. maintenance cost per unit output, maintenance cost over replacement value and maintenance cost over production cost).

Dwight classifies performance measures in a five-level hierarchy according to their implicit assumptions on the impact of the maintenance system on the business: overt (visible) bottom-line impact measurements (e.g. direct maintenance cost), profit-loss and visible cost impact measurements (e.g. total failure/down time cost), instantaneous effectiveness measures (e.g. availability, OEE), system audits (e.g. % planned work and work backlogs) and time related performance measurements (e.g. life cycle costing and value based performance measurement). Dwight's work looks at the variations in lag between an action and its outcome.

It is clear that each author has a unique way to classify maintenance indicators. They also differ in their choice of indicators. However, some indicators and categories of indicators are recognized by all authors as vital for management of the maintenance function. For example, much emphasis has been placed on equipment performance in terms of number/frequency of breakdowns, MTTF, availability and OEE. Similarly, maintenance cost-related measures are deemed important. Measures of maintenance efforts are considered important by many authors, though they use a variety of terminologies to describe them (e.g. maintenance productivity and operational purpose fullness, maintenance efforts, maintenance work management. Interestingly, while the literature proposes common lists of KPIs, it lacks an agreed-upon methodological approach to select or derive them. As a result, users are left to decide the relevant KPIs for their situation. Given the lack of consensus, one of the objectives in this paper is to investigate how maintenance KPIs is chosen.

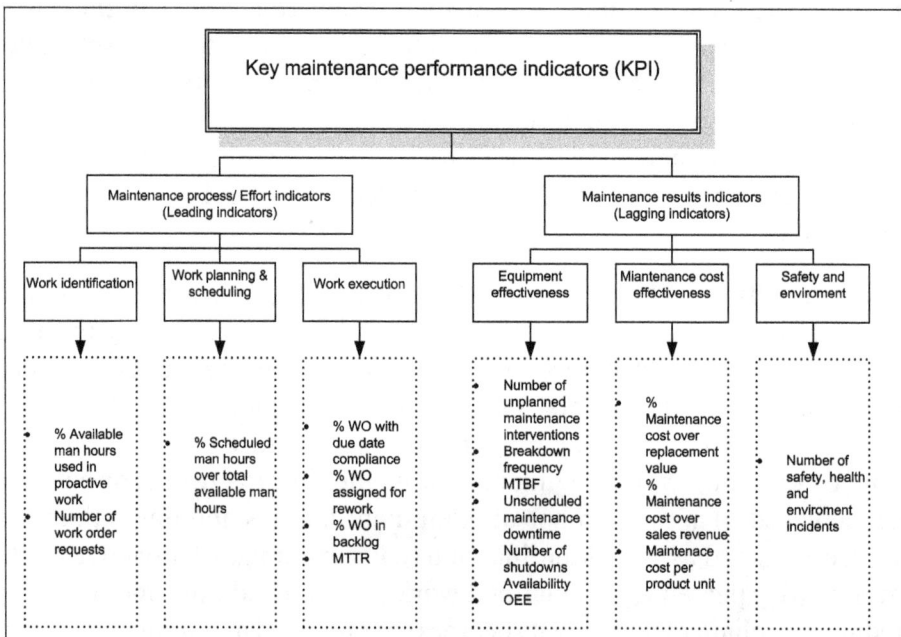

Key maintenance performance indicators in the literature.

Based on the literature, the commonly used maintenance performance indicators fall into two major categories. The maintenance process or effort indicators are defined as leading indicators and the maintenance results indicators as lagging indicators. Using the definition of Weber and Thomas, leading indicators monitor whether the tasks being performed will lead to the expected output and lagging indicators monitor the outputs that have been achieved. Under maintenance process indicators and according to Muchiri et al. there are three categories of indicators: work identification, work planning and scheduling, and work execution indicators. For maintenance results, there are three categories of indicators: equipment performance, maintenance costs and safety and environment indicators. Each category has its own performance indicators.

The survey objectives include investigating the extent to which these indicators are used in industries, establishing which are most frequently used, i.e. popular indicators, and determining how effectively they are used in maintenance management.

Grouping Indicators: Frameworks and Scorecards

For the most part, the focus has been put on performance measuring systems rather than on individual performance indicators. An overview of the most commonly used performance measurement systems, with their respective advantages and disadvantages appears below. The systems discussed differ by the choice of the indicators and the manner of representation.

- Global PIs: In practice, maintenance performance is often judged on a single indicator value. More frequently, however, a complex ratio is used, in which a number of relevant factors are combined, sometimes with different weights. A typical ratio is: yearly costs for materials, labor and subcontracting/yearly budget. This is a popular concept because of its compactness. A ratio is tricky to use because of the strong aggregation which may cause cancellation of some effects (e.g. an increase in labor cost and a comparable decrease in materials cost will never be apparent in the indicator), making it difficult to know what exactly happened (did all costs increase, or only one, or some, etc.?).

- Set of PIs: A number of PIs are used, each highlighting an aspect of the maintenance activities. For example, for maintenance stock, the following indicators are often used: inventory value, number of items in stock, turnover, number of new/obsolete items and number of in/out movements. This gives a more complete view of maintenance performance, but does not always allow a clear evaluation because of the lack of a structured framework.

- Structured list of PIs: Various aspects of maintenance activities are evaluated at the same time; for each aspect, a different set of indicators is used. The TPM measures, evaluating the well-known six big losses in equipment performance, may be considered a special type in this class.

The most popular set or list of indicators is a scorecard. The Balanced Scorecard (BSC) is frequently used to group maintenance KPIs and show different faces of the maintenance function. The balanced scorecard is a holistic approach which groups both financial and non-financial measures to measure performance. In any organization, the corporate objectives state the company's vision. A corporate strategy is formulated as the way to achieve these objectives. A corporate balanced scorecard is part of the corporate strategy to measure performance and compare it with the corporate objectives. These balanced scorecards are applied to different divisions and departments, right down to the employee level.

Similarly, maintenance performance indicators can be translated from balanced scorecard perspectives and applied to the divisions, departments, sections and employee levels. The maintenance objectives are linked to critical success factors, key result areas, and key performance indicators. The critical success factors support the maintenance objectives. The key result areas are the success areas where the key factors can help achieve the maintenance objectives.

Daryl Mather has adapted the Maintenance Scorecard, MSC from the original balanced scorecard for asset-centric industries such as electricity generation and distribution, water treatment, oil and gas, mining, rail, and heavy manufacturing. Based on RCM2 by John Moubray and his interpretation of functional measurement and monitoring machine performance, as well as Kaplan and Norton's traditional BSC approach, MSC identifies a need for strategic initiatives and determining how to best determine what form of intervention is required. Breaking down indicators from the corporate levels of management is a common practice, and it has been applied effectively on the front lines of maintenance activity.

Mather proposes using the MSC approach to develop and implement strategy in the area of asset management. This will help identify strategic improvement initiatives and the areas they focus on, early in the process. The MSC is a methodology based in the measurement of performance, built around the use of management indicators, which can lead the development and implementation of a strategy.

A different approach to measuring need is given by they argue that measuring performance by means of scoreboards focuses on the process safety and environment as a necessary consequence of maintenance activities. These activities are important for plant safety. According to the authors, quality management systems, risk prevention and safety require the implementation of a metric in the maintenance department.

Hierarchy of Indicators

Indicators are commonly formulated at different levels as well. Each level serves certain purposes for specific users. Users at the highest level of the management traditionally refer to aspects that affect firm performance, whereas those at the functional level deal with the physical condition of assets. The use of multiple performance measures at the level of systems and subsystems helps to solve problems. If a corporate indicator shows a problem, then the next lower level of indicators should define and clarify the cause of the weakness that has caused this problem.

According to Mitchell, a hierarchy of different parameters, all linked to business goals, is vital for the success of a program to manage corporate physical assets.

Many authors agree that multifaceted maintenance requires metrics, and those metrics should serve specific levels of the organization's hierarchies. TRADE shows the levels of performance indicators in a typical organization, figure shows different organizations have different hierarchies of performance measurements.

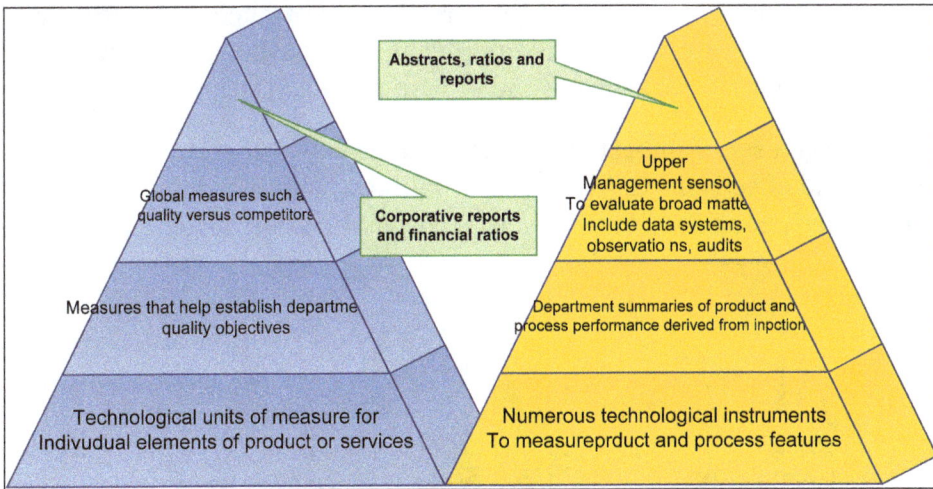

Pyramid used in all levels of the company, TRADE
Units of measures to the left and sensors to the right.

Grenčík and Legat, like TRADE, make analyze the consistency of the indicators and their management classification levels. To select the relevant indicators, the first step is to define the objectives at each level of the company. At the company level, the requirement is to determine how to manage maintenance to improve overall performance (profits, market shares, competition, etc). At the level of production, performance factors which have been identified through prior analysis are more important; these include improved availability, improved intervention costs, safety, environmental preservation, improvements in maintenance costs, value inventory, contracted services control, etc.

Kahn suggests using Key Performance Indicators (KPI) to set up a hierarchical methodology to quantify project improvements in the maintenance function. He suggests that to visualize the expected benefits, the process variations and trends should be adequately monitored. The established KPIs should be controlled and an adequate program for continuous improvement should be set up. For Kahn, a KPI is a traceable process metrics that allows decision making aimed at established business objectives; maintenance KPIs should include indicators of corporate level, such as the

OEE, or financial, such as the overall maintenance budget compared to replacement cost, and so on. These financial indicators should be positive to ensure the organization's support for maintenance improvement projects. Like TRADE, Kahn proposes five levels of KPIs, each with its own requirements and target audience, thus consolidating the segmentation of indicators by levels of organization. The five levels include: Maintenance costs; Availability of equipment; OEE; Production costs; Performance.

Campbell classifies performance measures into three categories based on equipment performance measure, cost measures and process performance measures. Kutucuoglu et al. suggest another general classification for a balanced performance measure. Their five proposed categories include: equipment related performance measures; task related performance measures; cost related performance measures; immediate customer impact related performance; learning and growth related measures.

Wireman defines a set of indicators divided into groups: (a) corporate, (b) financial, (c) efficiency and effectiveness, (d) tactical (e) functional performance. The indicators should be properly connected to the levels of corporate vision and company mission.

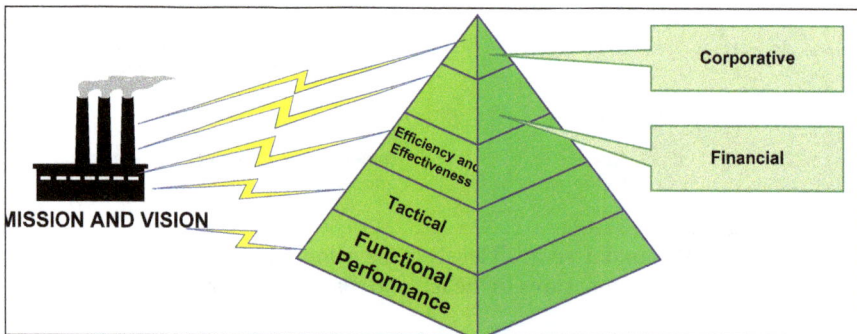

Hierarchy of indicators in maintenance.

These concepts all suffer from a hierarchy that condemns low levels to work with operational and functional indicators, while assigning economic indicators to top management, thereby dividing the analysis and creating indicators of first and second categories. Most authors have traditionally associated maintenance metrics with RAMS parameters, but these are only part of the performance to be measured. A few have included cost, and a few more have integrated a number of corporate indicators into the maintenance function.

These groups and hierarchies of PIs are ambiguous and nonuser defined. In fact, they confuse groups with organizational levels. There is no end user identification and no attempt to have responsible people involved in continuous improvement actions. In implementation, therefore, there should be multilevel indicators. According to Wireman, the first layer could be at the corporate strategic level; the supporting level could be the financial PI; the third would be efficiency and effectiveness indicators, with fourth and fifth levels of tactical PIs and functional PIs respectively. The five levels of the pyramid show the hierarchical relationship of the PIs. It should be noted that the indicators are always determined from top down, using corporate indicators measures; what is important to top management is to satisfy the needs of the stakeholders/shareholders.

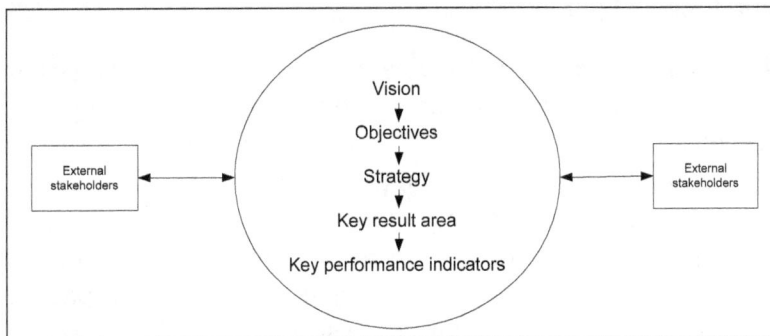

Developing Performance Measurement
Indicators from Vision, Objectives and Strategy.

For Parida, indicators in three levels must be considered from the perspective of the multi-hierarchical levels of the organization. The first hierarchical level could correspond to the corporate or strategic level, the second to the tactical or managerial level, and the third to the functional/operational level. There could be more hierarchical levels depending on the organizational structure.

These corporate PIs will vary from company to company, depending on current market conditions, the business life cycle, the company's financial standing etc. Thus, PIs must be tied to the long-range corporate business objectives of a specific company, meeting the needs of both the operations and the maintenance processes. The critical strategic areas vary from company to company, and from sector to sector, but generally include areas such as financial or cost related issues, health, safety and environment related issues, processes-related issues, maintenance task related issues, and learning, growth and innovation related issues. They combine the internal and external concerns of the company.

The measurement system should cover all processes in the organization. There must be a logical interconnection between indicators, so that the numbers can be interpreted and a good conclusion can be reached, thereby allowing good decision making. This premise implies a hierarchy of indicators addressed in a dual way, Caceres. Maintenance indicators will be segmented according to the organization's areas of influence, due to the interactions of the maintenance department with finance, human resources, purchasing and, of course, production to achieve corporate objectives. Simultaneously, these indicators correspond to different levels in the organizational structure, so they will be targeted to them.

For Cáceres, performance measurement must be comprehensive and requires an appropriate scorecard. He argues that management should be measured holistically, not only in the financial perspective as is traditional, APQC. Maintenance performance should be based on maintenance parameters of availability, reliability, mean time to repair. In addition all perspectives within maintenance should be integrated to cover organizational and technological aspects, internal processes, customer and company perspectives and financial perspectives.

Bivona and Montemaggiore agree with Cáceres and argue there is a lack of linkage between the objectives of general maintenance and the business strategy adopted because of performance measurement systems. The most common performance indicators oversee operational management from the unique perspective of the maintenance activity, ignoring the effects of maintenance policies on company performance and their impact on other departments. Some authors argue that the performance measurement system based on relations between different departments of a company facilitates the communication process between the corporate strategy and the various hierarchies of the maintenance organization.

This leads to an alignment between business objectives and maintenance. To this end, most authors suggest adopting the balanced scorecard approach to the formulation of maintenance strategies, not only as grouping of indicators but also as a hierarchy. The systemic perspective of the balanced scorecard, in fact, supports management in analyzing the various relationships between the subsystem maintenance and other business areas, to prevent the gains or losses in the performance of maintenance management that are included in the execution costs of other departments.

The balanced scorecard method first developed by Kaplan and Norton later adapted by Tsang et al. for measuring maintenance performance is an effective way to measure maintenance performance. It designs the maintenance performance measure using the following four perspectives:

- Financial (the investor's view).

- Customers (the performance attributes valued by customers).

- Internal processes (the long-term and short-term means to achieve financial and customer objectives).

- Learning and growth.

This technique can link the maintenance strategy with the overall business strategy and develop performance measures for maintenance that are linked to the organization's success.

Alsyouf criticizes the balanced scorecard technique suggested by Tsang et al, arguing that the performance measures based on the four non-hierarchized perspectives are top-down performance measurements which do not take into account the extended value chain, i.e. it ignores suppliers, employees and other stakeholders. The extended balanced scorecard presented by incorporates performance measures based on 7 perspectives: corporate business (financial), society, consumer, production, support functions, human resources and supplier perspectives, as shown in figure below.

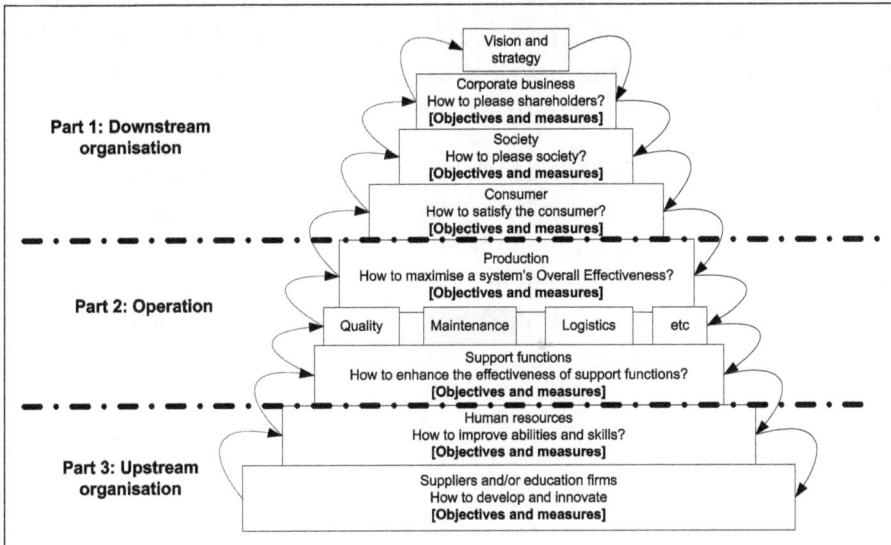

In any planning and development activity, offers several alternatives, and one must choose the best-fit. Normally, the objectives of the decision maker are reflected in various criteria. If there are a number of criteria, multi-criteria choice problems arise; this is solved by having information on the relative importance of the criteria. The selection of factors or variables constituting various performance criteria, such as productivity, effectiveness, efficiency etc., is an important step in developing a performance measurement system in an organization. This is conceived essentially as multi-criteria decision making.

In an MPM system, a number of criteria or goal functions must be considered from different stakeholders' points of view. These criteria can be broken down into maintenance indicators, such as mean time between failure, downtime, and maintenance cost, planned and unplanned maintenance tasks, etc.

The operational and strategic levels of these maintenance indicators need to be integrated as well. The development and identification of MPIs for an organization consider the company's vision, objectives and strategy, as well as the requirements of external and internal stakeholders, as given in figure. In our development of a MPM framework, we consider the basic four perspectives of Kaplan and Norton's balanced scorecard, along with the maintenance criteria. In addition, we consider health, safety, security and environment and employee satisfaction to make this MPM system balanced and holistic from the organizational point of view.

Multi-hierarchical Levels

MPIs must be considered from the perspectives of the multi-hierarchical levels of the organization. The first hierarchical level could correspond to the corporate or strategic level, the second to the tactical or managerial level, and the third to the functional/operational level. Depending on the organizational structure, there could be more than three hierarchical levels. The maintenance indicators on the functional level are integrated and linked to the tactical or middle level to help management in its decision making at the strategic or tactical level. It can be challenging to integrate MPIs from top-down and bottom-up flows of information.

Involving all employees in this MPI development process is another challenge. So that everyone speaks the same language, the strategic goals need to be broken down into objective targets for operating maintenance managers, and this which may act as a performance driver for the whole maintenance group. Thus, the objective output from the operating level in terms of KPIs is linked to strategic goals; moreover, the subjectivity increases as the objective outputs are integrated with KPIs at higher or strategic levels.

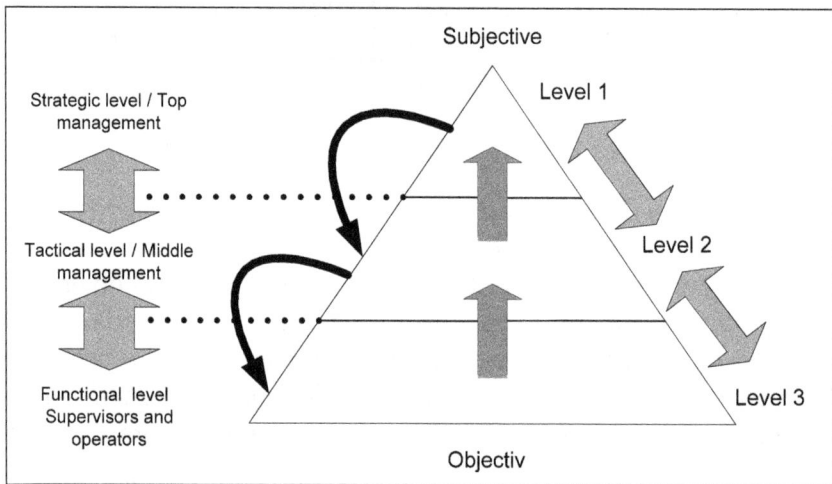

Hierarchy levels of MPM model.

MPM Frameworks

The MPM framework is a vital and integrated part of the PM system of organizations. The need to develop and implement an MPM framework is well established. The MPM framework must link performance measurements with the organizational strategy and consider criteria consisting of financial and non-financial indicators. However, little literature covers the development of a systematic approach to PM in maintenance; even less embraces every level of maintenance. A possible exception is work by Kutucuoglu et al.

The development of a MPM system is intimately linked with the overall PM system and the overall corporate strategy. Therefore, it is imperative to look into the shortcomings of PM systems, especially those systems based on financial measures only.

Tsang et al. says that a US survey of 200 companies in 1995 concluded that despite reasonably high level use, non- financial measures and targets are frequently treated in isolation from strategic objectives. They are not reviewed regularly, nor are they linked to short-term or action plans; rather, they are largely ignored or for interest only.

In another study, Tsang et al. looked at a maintenance organization using a structured process to identify performance measures. They found that the management was not aware that a PM system could achieve vertical alignment and horizontal integration of activities across organizational units. Thus, performance measures were primarily used for operational control only.

Coetzee provides a comprehensive list of MPIs and ratios and identifies 21 indices under the four categories of machine/facility maintenance efficiency, task efficiency, organizational efficiency and profit/cost efficiency. The MPIs are set within a hierarchy, but Coetzee fails to identify the specific hierarchy in the organization who uses them. Nor are these MPIs clearly connected to the corporate strategy. Riis et al. design a framework showing cross levels and the functional integration of maintenance management which attempts to relate maintenance to manufacturing strategy. However, it does not take care of other customers and suppliers, such as design, finance, top management and issues including health, safety, security and environment (HSSE), employee, and corporate strategy.

Kutucuoglu et al. and Sharp et al. adapt TPM and TQM to improve maintenance performance and identify critical success factors (CSFs) linked to maintenance. Dwight explains two other approaches; 'the system audit approach' and the 'incident evaluation approach', which defines performance in terms of changes in value systems. Value is defined here as the probable future earnings of a system.

Tsang and Ahlmann adapt the balanced scorecard of Kaplan and Norton to create a strategic approach to MPM. However, Kaplan and Norton are limited in that they do not consider employee, societal or environmental aspects.

Kutucuoglu et al. develop a performance measurement system for maintenance using the matrix structure of the quality function deployment (QFD) method. Murthy et al. present the Strategic Maintenance Management (SMM) approach which has two key elements:

1. Maintenance management is a vital core business activity, crucial for business survival and success, and as such, it must be managed strategically.

2. Effective maintenance management needs to be based on quantitative business models that integrate maintenance with other decisions, such as production etc.

The multi-disciplinary activity of SMM involves data collection and analysis to assess the performance and state of equipment, building quantitative models to predict maintenance and operation impact on equipment degradation and managing maintenance from a strategic perspective. This approach is not balanced and integrated, as it does not consider all stakeholders.

In a project for the Norwegian oil and gas industry, Ellingsen et al. suggest a PM framework based on a balanced scorecard model and a list of key indicators. The framework considers cost, operation, health, safety and environment and organization perspectives. Maintenance and employee satisfaction are not included.

Categorization of Indicators

Financial Indicators

Financial measures are often considered the top layer in the hierarchy of the measurement system used regularly by senior management. This level seeks to achieve a good return on its assets and to create value and its metrics are used for strategic planning. Therefore, the metrics at this level are the backbone of the organization. This level can also be used to compare the performance of departments and divisions within the parent organization.

The financial figures are lag indicators and are better at measuring the consequences of yesterday's decisions than predicting tomorrow's performance. To overcome the shortcomings of lag indicators, customer oriented measures like response time, service commitments and customer satisfaction have been proposed to serve as lead indicators. Examples of such measures are the return on investment (ROI), return on assets (ROA), the maintenance cost per unit of product, the total maintenance costs in relation to manufacturing costs, etc.

Vergara proposes the net present value (NPV) for use in maintenance as one of the financial indicators. NPV consists of knowing how much could be gained from an investment if all income and expenses could be made instantly. Therefore, NPV is used to determine whether an investment is appropriate. It is used in many sectors and areas of the company but rarely in maintenance. Tsang et al. presents a better performance measurement technique first proposed by Dwight which takes into account the impact of maintenance activities on the future values of the organization, instead of concentrating on short term values. However this technique also concentrates only on the financial dimension of performance measurement and is quite laborious to implement.

Hansson proposes using a battery of financial indicators to study the results found for maintenance departments. He suggests creating a proper benchmarking of the maintenance function, arguing that one should consider such measures as: the percentage change in sales, return on assets, return on sales, the percentage change in total assets and the percentage change in the number of employees. These are generally accepted as indicators of financial results and facilitate comparison of results with other studies, e.g. Hendricks and Singhal. The correlation of such indicators with tactical maintenance indicators links maintenance with corporate strategy.

Coelo and Brito propose incorporating financial indicators into maintenance management. This hypothesis confirms the importance of a system for measuring enterprise performance based on indicators, with special emphasis on maintenance. Coelo and Brito discuss the need for integration to achieve a harmonious balance of financial performance indicators and the strategic vision of maintenance efficiency.

Cáceres argues that all planning systems should show the history of strategy and corporate positioning indicated in its financial goals, linking them to the sequence of actions to be undertaken with customers, internal processes and finally with the employees themselves. This perspective is focused on return on investment, added value to the organization and reduced unit costs. In maintenance, the costs of each activity the incidence rate of maintenance costs per unit of production and maintenance costs on the value of assets are monitored to indicate the company's global position.

Indicators Related to Human Resources

The adoption of a separate category of measures relating to human resources reveals the uniqueness of maintenance services. Maintenance organizations depend entirely on the performance of employees to achieve their objectives. But the quality of employees' work in the maintenance services cannot be measured directly. Knowing their experience, education, training and skills is essential to adequately measure the result of work done. Few organizations measure the excellence of their human factor; nor do they include this factor in their assessment of the performance of the maintenance function. In addition, measures of organizational performance are often selected on the basis of convenience. Typically, these measures are either too narrowly or too broadly defined, Cameron. Measures include indirect/direct labor, labor in reserve, training measures, and percentage of overtime.

To Cáceres, the introduction of KPIs to maintenance human resources should cover what he calls the organizational dynamics perspective, where excellence focuses on the people and the culture of the organization, identifying the key skills that support internal targets. Ultimately, this is a true reflection of the labor climate in the microclimate of maintenance, Rensen. A measure related to human resources is the company's intellectual capital (IC). Fernandez and Salvador discuss the importance of intellectual capital in maintenance teams, noting that it has become a critical issue for many organizations. They incorporate key indicators related to this intangible aspect into their balanced scorecard.

In the area of human resources in maintenance, special attention must be paid to the prevention of labor hazards. For this reason, a number of authors propose an indicator of equipment safety. Many operators in maintenance areas are affected by workplace accidents, far more so than production workers, Manuele. The maintenance staff is more exposed to such high risk factors as electric shocks, the dropping of heavy components, contact with chemicals, etc. For production staff in general an accident is due to the failure of accident prevention measures or the breaking of established procedures. In any case, whatever the origin, an accident has negative effects on employee morale, stops production and affects the reliability of equipment. Guaranteeing safe equipment and a safe environment, as well as cooperating with regulatory agencies, is a maintenance function. Maintenance must be rigorous in developing and enforcing security procedures, and in maintaining barriers to prevent accidents.

Indicators Relating to Internal Processes of the Department

Some authors refer to indicators of internal procedures as functional measures. Traditionally, this category includes processes related to efficiency that are measured within a maintenance organization. Examples of these processes are: work orders, inventory and purchasing, and management information.

For Cáceres the internal KPI perspective, or the process perspective, is related to the work process and to improvements in the pursuit of excellence. The purpose is to understand the processes that add value to the business and identify the drivers of domestic objectives. In the specific case of maintenance management, indicators are usually set as repair times, overtime, certified processes, security aspects in the activities and the implementation of plans and programs. This perspective includes the measurement of the internal mechanics used for the proper development of other perspectives.

Technical Indicators

Some authors refer to this category as the technical level of performance indicators. The first objective of this set of indicators is to measure equipment (assets) performance, at least that equipment considered part of the maintenance function. Mitchell says: 'At the technical level, the figures are used to monitor the performance of certain processes, systems, equipment and components'. This level is concerned with the effectiveness of maintenance work.

Functional Safety as a Key Indicator for the Client

Cea proposes the overall indicator of functional safety as KPI for the customer; it is what he or she expects from the assets. For Cea, for functional safety to be achieved, the user must receive the service that he or she expects from the system, with established quality and safety standards being met.

According to Blanchard et al. functional safety is 'the probability of the system to complete its mission, since the system was available at the beginning of the mission'. Functional safety is a measure of the system condition at one or more points during the mission; it is strongly influenced by the system reliability and maintainability and the quality of the project Reliability is associated with the compliance function over time and system performance. Maintainability is associated with the ability of the equipment to recover function when it stops, Kumar, Castro and Lucchesi Knezevic et al.

Therefore, according to Blanchard, direct and indirect indices of functional safety are the availability, reliability, maintainability and safety of a production system. This system must have an information subsystem based on indicators of efficiency and feed-back; it must be a valid tool operationally, whereby the user can fully appreciate the benefits of having a 'safe operating system'. Indices are reflected in the operational performance of assets and the quality of the products produced. For Cea, RAMS parameters are the basic components of the key indicator, i.e. functional safety. Söderholm as well as Cea, refers to functional safety as the basis of the whole system of indicators for maintenance and to RAMS parameters as the primary indicators on which to build the entire scorecard.

Changing Role of RAMS Parameters

For a long time, RAMS parameters were the only indicators adopted for measuring the performance of maintenance according to the purely technical or operational aspect assigned to this function. Currently, they have a more privileged role, albeit limited to the quality of the service that the maintenance function gives to its customers.

Killet emphasizes the practice of maintenance performance indicators focused exclusively on operational aspects. Without underestimating these indicators, Killet notes the need to develop corporate indicators. In fact, many authors have recently expressed concern over the limiting of maintenance indicators and scorecards to operational aspects that may be important for the consideration of client-related issues, but lack the broader vision of the maintenance function within a company Geraerds.

Martorell et al. agree with Blanchard that there are two main categories of indicators. The first category includes the direct or basic indicators that are directly linked to the collection of operational data, such as the number of maintenance actions, the type of maintenance activities, and cost. The second category includes indirect indicators, derived from direct indicators, such as reliability and availability. For Martorell et al. the direct indicators can be linked to functional requirements, while the indirect indicators can be related to non-functional ones.

The categories are technical, economic and organizational indicators. These levels do not correspond to organizational hierarchies, but represent a progressive decomposition of higher level indicators into more specific indicators.

The classification into technical, financial and organizational indicators refers to aspects of effectiveness (time-consuming task) and efficiency (cost and manpower involved). Effectiveness falls within the realm of technical indicators, while efficiency is financial and organizational. Organizational indices, due to the high importance of human factors in the maintenance function, play an independent role. However, some authors consider them attached to costs and part of efficiency. All authors see the RAMS parameters as raw materials for creating more complex indicators of effectiveness, thus providing the maintenance scorecard with more indicators of efficiency.

Presentation of Performance Measures

One of the most important factors for a successful performance measurement system is a good introduction of the indicators. If they are not presented and explained to users, they may be inappropriately used. Mitchell says that 'to fully exploit the benefits of metrics, the metrics must be clearly visible'. Seeing the figures often has a positive effect, encouraging everyone to achieve the objectives in the functional area being measured.

Kaydos states that having the performance measures visible to everyone has two advantages. First, everyone can be proud of what has been achieved. Second, where nothing has been achieved, the pressure exerted by workers in other departments has a positive impact. There are a variety of ways to present a performance indicator, depending on the type of information needed and the type of user. Charts, graphs, figures or just numbers can deliver a performance indicator.

Lehtinen and Wahlström emphasize the visual aspect and simplicity of the indicators, because this will be a key in developing a subsequent metric. The indicator should stand out in reporting and promoting an atmosphere of reports with proper quantification and calibration of the problem. It is believed that this environment is an important part of continuous learning from which excellence is achieved. Lehtinen and Wahlström show the nature of different indicators and the need to present them in a visually attractive and powerful way for workers involved in maintenance and production processes.

Another issue associated with the presentation of performance indicators is the frequency of their presentation. Some indicators require continuous data collection; others may have a monthly frequency. There is no advantage to measuring more frequently than needed, as it only increases costs.

Today, technology allows the use of an online graphical user interface (GUI) for presenting and monitoring indicators tailored to each person's personal needs, thereby making it possible to use the same system for presentations throughout the hierarchy. Notices can be sent automatically to mobile devices or mailboxes to further increase information efficiency.

Efficiency of Performance Measures

Metrics must be understandable, addressed to the needs of users and controllable by managers through work activities, Mitchell. According to Kutucuoglu et al., to develop an efficient and effective system of performance measurement, it should be clear what indicators are to be measured, how to do it, the timing, and who should implement the measurement. In fact, for Manoochehri, three obstacles to the effective development of metrics are the misidentification of indicators, less than rigorous data collection and the bad use of indices by the company managers.

According to Besterfield et al. to be useful, measurements should be simple, understandable, and few in numbers so that users can concentrate on those that are most important to them.

1. The number of indicators to be included and their origin in the adopted metrics: Killet reflects on the number of indicators to include and the property of each based on studies by Woodhouse 2000, and in line with the characteristics previously proposed by Besterfield. Woodhouse argues that the human brain can only handle four to eight measurements intended to quantify the goodness of one aspect. This suggests that it would be reasonable to target a maximum of six measurements for each supervisor/manager. To achieve this objective, he proposes the measurement of key characteristics, limiting the amount of information used and the sources from which to extract this information. In a multifunctional organization, it is likely that other departments may collect and share some of the data. For example, the collection of data on availability and reliability can be relatively simple for the production department. The department of labor risk prevention is ideal for monitoring injury rates, and the human resources staff is better situated to provide data on absenteeism. This supports Besterfield's thesis on ownership of data.

   Shreve in line with Woodhouse 2000 and with respect to specific indicators of Condition Based Monitoring (CBM), proposes the selection of six to eight indicators of high-level performance to analyze the effects of a CBM program in a factory. The performance indicators can be used both in production and in maintenance to display the program progress. The parameters for monitoring the results of the CBM should be established before its implementation.

   The author further emphasizes that measurements should be directed towards areas with the greatest impact on improving, ignoring those with a small ROI. Without constant reminders of performance, programs can start strong, but then rest on the initial achievements without ever reaching maturity. Proactive measurements should be the goal of the monitoring program. This level of maturity is based on the desire to find any problems affecting production rates and product quality before they appear. Shreve and Woodhouse agree that the goal is to find indicators with the highest ratio of implementation impact at each level instead of short-term self-satisfaction results.

2. Data accuracy: The performance model is expected to give the correct output result since the right data is fed into the model. The model must be accurate and consistent when processing the input data. This implies that the processing capability of the model must strong enough to accept the required data input and release the correct information to achieve a particular task.

   It is clear that 'garbage in, garbage out' holds for any sequential-factored data system model. For the result of model evaluation to be correct, the input data must be correct. To avoid error right from the start, the model user must ensure the correctness of the data input and, therefore, of data computed. The values must be checked for homogeneity in units, readings, computations, etc. The introduction of incorrect or corrupt data in the performance measurement system is harmful, leading to wrong decisions and losses. Thus, indicators for data accuracy monitoring are necessary. However, a good performance measurement system does not require high precision, Kaydos. In most cases, one needs to know how to identify problems. Very accurate performance measurement is not required; it is more important to know if the trend is up or down and to know how its current value compares to historical measures. If the way an indicator is calculated is changed, it is crucial to overlap so that the trend is not lost. Kaydos also stresses the importance of trust and credibility: if users do not trust the organization to generate the proper measures, the whole system is useless.

Barringer and Weber say that frequently the data available to exploit are sparse, poorly collected or of doubtful veracity. Barringer and Weber suggest that understanding how to manage the reliability of data is the first step towards solving the problems. For Tavares, indicators like MTBF or MTTR are particularly accurate. Their high level of accuracy is linked to the number of items observed and the observation period. The more records that are available, the greater the accuracy of the expectation values are. In the absence of a high number of items, or if one wishes to obtain the average time between failures of each one separately, according to Tavares, a fairly extensive study (five years or more) is advisable.

3. Metrics understanding by users: Users must be able to assess or evaluate the performance of the MPM system in terms of its operation and results. More importantly, the user must know how to assess, analyze and interpret the end result of computations to discover knowledge embedded in the operation of the MPM system. That is why user training is so important. The user must have the knowledge necessary to use a performance measurement system. It is assumed that the user has taken courses in areas such as mathematics, physics, programming, statistics, etc. to understand the model's procedure and application. As part of the training, the user must be assessed to determine the level of competence attained.

In Manoochehri, the effective use of performance measures requires user education because a misunderstanding leads to wrong decisions. Major problems that could lead to a failure in the measurement of system performance are a lack of real management commitment, a lack of attention to business goals and objectives and an incorrect updating of performance measures.

The failure to use performance measurements in an organization may be the result of not overcoming the challenges associated with the implementation of a new set of performance indicators. It is, therefore, very important for the implementation team to concentrate on the project, especially at the beginning. If this is not done, it could result in a loss of confidence in the new system and a lack of voluntary participation in its development.

The performance measures of the system must be designed to serve the purpose of the organization. According to Wireman, multiple indicators should be associated with every level. One layer of indicators could be at the corporate level, and another at the departmental level. The levels may vary depending on company size.

Furthermore, to successfully implement a performance measurement system, measurements should not be numerous. Dispersion into too many areas simultaneously can lead to information overload, making it more difficult to direct limited resources to higher value activities.

A challenge faced by most performance measurement systems is change. But this is part of the manufacturing business. A measurement system should not be affected by changes in production characteristics but it must be adapted to them. Moreover, indicators may become out of date.

Maintenance Audit

Converting forecasts into concrete real numeric values require extraordinary effort. If this proves too onerous, a company may choose to focus on the system and its attributes rather than on specific outcomes using what can be termed the 'system audit approach'. A maintenance audit is an examination of the maintenance system to verify that maintenance management is carrying out its mission, meeting its goals and objectives, following proper procedures and managing resources efficiently and effectively.

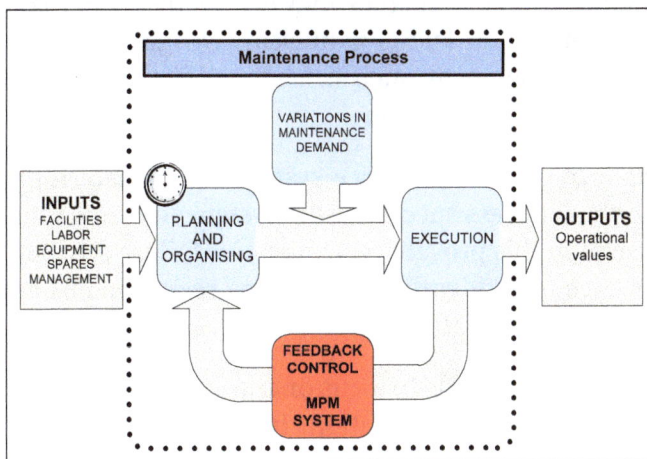

Input-process-output model (IPO model) of the maintenance process.

This concentrates on the maintenance system itself, as opposed to quantifying its inputs and outputs. The results obtained from this approach should have a level of accuracy that is compatible with the information normally available about real performance. Subjectivity in performance measurement will not be overcome, but such subjectivity will be made more visible.

Auditing, as a general technique, can be divided into two categories. The first utilizes general audits based on a common assumed standard designating what constitutes a good system. This is a popular tool for consultants since it allows them to have a consistent standard of what a good maintenance system should be. This is normally isolated from a deep understanding of the subject organization's business. It permits consultants to insert important attributes of which they have a good knowledge, but whose importance varies in fact from one organization to another. This kind of audit is a thorough review of the various dimensions in the maintenance system, including organization, personnel training, planning and scheduling, data collection and analysis, control mechanisms, measurements and reward systems, etc. To get unbiased findings, the reviewer should have no direct responsibility or accountability for the performance of the system under review. The audit is usually conducted by using a questionnaire designed to provide a profile of the maintenance system.

Typically, the questionnaire is structured to address specific key areas in the system to be audited. Responses to these questionnaires may take one of these forms:

- Either yes or no.

- Choose one or more of the available options.

- On a Likert-type scale of, as an example, 1 to 5, to indicate different degrees of agreement or lack of it.

Different weights may be assigned to questions to reflect their relative contributions to the system performance. Even though they may use sophisticated assessment schemes, the underlying theory of system audits is unclear. Dwight suggests a procedure that relates the state of a system element, such as feedback from operations, to its contribution to the system's overall performance.

The overall performance of a maintenance system can be determined by aggregating the contributions to the business success of the system elements that influence the relevant failures of assets. In this procedure, exhaustive failure attributes that contribute to business success have to be identified. The same requirements apply to the system elements that influence a failure attribute.

The more typical system audit tends to focus on the issue of conformance to a standard model, both in system design and execution. It is assumed that the standard can be universally applied to achieve superior performance. The maintenance system audit questionnaires by Westerkamp and Wireman rely on this concept. This approach to system audits fails to recognize that different organizations operate in different environments. Products, technology, organizational culture and the external environment are key variables in an organization's operating environment, and they may be in a state of constant change. Superior performance can be achieved only if the internal states and processes of the organization fit perfectly in the specific operating environment. Sociotechnical Systems (STS) analysis provides a methodology to design a system that will achieve this fit. Thus, the basic assumption of a standard reference model implicit in the design of the typical audit questionnaire is problematic.

The second category of audit technique is initially concerned with the analysis of technology and business constraints. This allows the determination of the relative importance, and required attributes, of the various elements of a system. The actual system attributes can then be analyzed against the ideal system and tempered by the requirements for excellence in the particular activities making up the system. This second technique is pursued here. It tends to be qualitative in its methods, as it seeks to quantify the judgments of people with knowledge of the maintenance system, the organization's requirements and the system elements to measure performance. Although this implies that it falls short of an objective measure, a compromise is forced in order to create an objective measure.

A maintenance system audit is necessary for developing an improvement action plan. According to Kaiser, The Institute of Internal Auditors, it helps management to achieve the following:

1. Ensure that maintenance is carrying out its mission and meeting its objectives;

2. Establish a good organization structure;

3. Manage and control resources effectively;

4. Identify areas of problems and resolve them;

5. Improve maintenance performance;

6. Increase the quality of the work;

7. Automate and recommend information systems to increase effectiveness and productivity;

8. Develop the culture of continuous improvement.

The audit process is usually done on site. It reviews key elements in the following way interviewing key people in the organization; conducting site inspections; reviewing process flows and mapping of maintenance functions and control; reviewing relevant documentations; demonstrating systems applications; attending key meetings; obtaining answers to structured questionnaires; and validating plant, equipment and maintenance performance. The results of the interviews and the answers to the structured questionnaires are analyzed to formulate action plans for improvement.

Westerkamp has developed an audit scheme that covers fourteen factors contributing to maintenance productivity. He advocates automating the auditing process through, organization staffing and policy; management training; planner training; craft training; motivation; negotiation; management control; budget and cost; work order planning and scheduling; facilities, store, material and tool control; preventive maintenance and equipment history; engineering; work measurement; and data processing. He suggests obtaining information about these factors by using a set of questions about each.

Kaiser has developed a maintenance management audit that includes key factors in the process of maintenance management: organization; workload identification; work planning; work accomplishment and appraisal. Each component has six to eight factors. Using structured statements and weights, Kaiser obtains an overall score for the maintenance system. In brief, necessary improvements can be identified from the audit process.

Duffuaa and Raouf have conducted a study on continuous maintenance productivity improvement using a structured audit and have proposed a structured audit approach to improve maintenance systems. They include the following factors in their audit: organization and staffing; labor productivity; management training; planner training; craft training; motivation; management and budget control; work order planning and scheduling; facilities; supplies/stores, material and tool control; preventive maintenance and equipment history; engineering and condition monitoring; work measurement, incentives and information systems. They propose using the analytic hierarchy process (AHP) to determine factors' weight and to compute a maintenance audit index. They also suggest root cause analysis to develop an improvement action program.

Duffuaa and Bendaya propose the use of statistical process control tools to improve maintenance quality, and Raouf and Bendaya suggest employing a total maintenance management (TMM) framework. An important component of TMM is a structured audit.

DeGroote argues for a maintenance performance evaluation approach based on a quality audit and performance indicators of maintenance. The quality audit should be conducted in the following four stages: survey of the influencing parameters; analysis of collected data, conclusions and recommendations; derivation of improvement action plan; and justification of the proposed improvement plan based on cost-benefit. The evaluation should include the following five major factors: production equipment; organization and management of maintenance; material resources; human resources; and work environment.

Price Water House Coopers has developed a questionnaire to evaluate maintenance programs. The questionnaire includes ten factors: maintenance strategy; organization/human resources; employee empowerment; maintenance tactics; reliability analysis; performance measures/benchmarking; information technology; planning and scheduling; material management; and maintenance process reengineering. The questionnaire features several statements about each factor; each statement is given a score ranging from 0 to 4.

Al-Zahrani has reviewed audit programs and surveyed managers and engineers in government and private organizations in the Eastern Province of Saudi Arabia to assess the factors affecting maintenance management auditing, with the aim of developing a suitable auditing form for facilities maintenance. He proposes an audit form consisting of six main components: organization and human resources; work identification and performance measures; work planning and scheduling; work accomplishment; information technology and appraisal; and material management. Each component has six to eight factors that are relevant to the performance of the maintenance system.

In the literature, five structured audit programs for maintenance systems have been developed by. The audit programs consist of key elements in the maintenance systems that are examined through a set of statements or questions. Each statement or question has a score and a weight. Then based on the audit, a total weighted score is compiled and compared to an ideal score. Based on these scores, an action plan for improvement is formulated. The process is repeated periodically to ensure continuous improvement.

In addition to the balanced scorecard technique, Tsang et al. presents a Systems Audit technique, based on socio-technical systems analysis (STS) for predicting the future maintenance performance and a DEA (Data Envelopment Analysis) technique, a non-parametric quantitative approach to benchmarking organizational maintenance performance in comparison with competitors. Using the four stage quality audit approach, Groote defines performance indicators in terms of ratios rather than absolute terms to develop a system for maintenance performance.

Many authors argue the necessity of obtaining both qualitative and quantitative results. Clarke suggests an audit must contain maintenance radar (a spider Bells-Manson type of graphic) which imagines all the economic aspects, human etc. of maintenance. In his view, a product of the audit must be good operative and technical maintenance practices. Like many others, Tavares makes use of radar maintenance in audits to represent the different areas of maintenance influence and dependency. Many authors agree that these radars should be generated from massive surveys. Despite the reliability of the surveys, the numeric data of the systems are not included in the radars; rather, they become subject to a strong human factor.

More recently Galar et al. have developed an audit in this category. Their model proposes a mixture of qualitative aspects, received through questionnaires, and different weights of maintenance KPIs that should strongly correlate with the questionnaire answers. This model shows the relation between trends in questionnaires and indicators that validates the correlation or highlights the divergence, merging qualitative and quantitative measures.

Benefits of a Performance Measurement System

Kutucuoglu et al. take advantage of the Quality function deployment (QFD) technique, using its ease of implementation, its alignment of performance indicators with corporate strategy and its ability to hold both subjective and objective measures, to develop an effective performance measurement system for the maintenance function. Their MPM system incorporates the key features necessary for effective maintenance performance measurement, i.e. a balanced view of a maintenance system, cross functional structure, vertical alignment of performance indicators etc. The introduction of performance indicators can:

- Clarify the strategic objective.

- Link maintenance department objectives to core business processes.

- Focus actions on critical success factors.

- Keep track of trends in development.

- Identify possible solutions to problems.

Varied industrial sectors have benefited from the introduction of indicators in their maintenance departments, Espinoza, in his work on the aluminum industry in England, says that in an effort to gain a competitive advantage over its main rivals, this industry is using a series of maintenance performance indicators through which its effectiveness can be monitored continuously (reliability, availability and use of equipment). Effective maintenance, according to the author, will vary the ratio from unplanned to planned activities. Espinoza says that increased scheduled jobs in this industry indicate that the maintenance strategy is effective. Racoceanu and Zerhouni have developed scorecards with sets of indicators in the machine tool sector to increase competitiveness.

Once indicators have been embedded in the organizational hierarchy, the benefits obtained by the organization, according to Wireman, include the proper development, evolution and progression of the maintenance model. Wireman proposes a sequential implementation of steps to ensure all the functions necessary for the proper management of maintenance are in place: 1) preventive maintenance, 2) inventory and procurement, 3) the work order system, 4) CMMS system, 5) technical training and staff relationships, 6) staff involvement in operations, 7) predictive maintenance, 8) RCM, 9) TPM, 10) statistical optimization, and finally 11) continuous improvement. Wireman considers that each of these activities is a section in the maintenance management process, as expressed in the pyramid shown in figure. Wireman suggests that a preventive maintenance program should be implemented before moving to the next level of the pyramid – and so on, up the pyramid. Before considering the application of RCM, for example, predictive maintenance programs and involvement of staff in maintenance functions, appropriate work orders systems and management of maintenance resources are required. The involvement of operators and employees constitutes the next stage; TPM programs will help for that purpose. Finally, optimization techniques will complete the organizational structure necessary for continuous improvement.

It is a mistake to reorganize a department using one technique, and it is advisable to avoid consultants and companies who advocate a single technology as a solution to a problem. When the department is in a state of 'stagnation' with respect to ratios or indicators, there is a need for a drastic change in maintenance philosophy. Reengineering may be a possible solution, but it is worth knowing that while moving up in the maturity pyramid of maintenance evolution, the next level will be more complex shows that when employees and managers perform functions in a process-oriented environment, organizations can measure their performance and pay them based on the value they create. In organizations that have undergone major reengineering, contribution and performance become the main bases for compensation and, therefore, in most cases, they are a success. Concerns with quality guarantees and product reliability cause organizations to focus their decisions on the efficiency and quality of maintenance management. When the process of reengineering is applied in maintenance, it affects other processes in the organization, and fewer people are needed to achieve standards of quality and efficiency. Zaldivar says that with fewer managers, fewer administrative levels, and, consequently, a predominance of flat structures, stable performance and a qualitative jump in each technical-economic indicator are assured, bringing the organization closer to the top of the pyramid of figure.

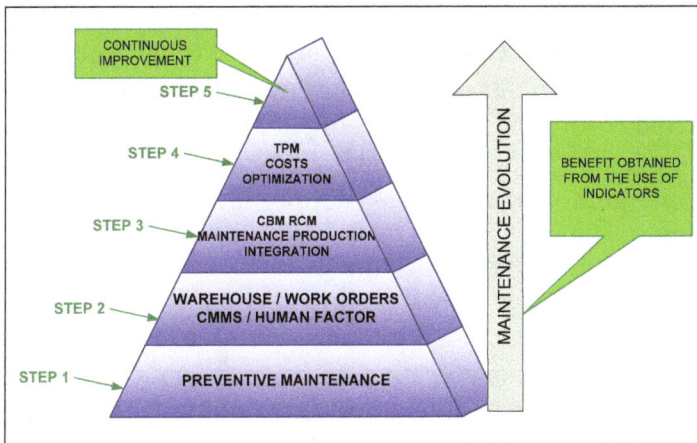

Construction of the maintenance management process.

This sequential development requires appropriate indicators associated with each level, from the operations related to each team, to the strategies of the organization at the top of the pyramid. The proper measurement of the state of maturity at each level will help to develop the next step. Unsystematic CBM implementations or isolated and uncoordinated strategies do not usually give good results. Proper progression through the pyramid is essential (as in the sequential development pyramid of comprehensive maintenance program and the proposed hierarchy of KPIs).

In his investigation of the ability of a maintenance process to reach maturity by climbing the pyramid mentioned above, Schneidewind notes the need for stability to ensure product quality. The use of metrics in the maintenance process predicts reliability and risk, thereby ensuring stability. Schneidewind stresses the need for strengthening each layer of the organization, as greater maturity will facilitate the jump to the next level. The evolution of the maintenance process should not be precipitated or accelerated.

E-maintenance

Today's competitive manufacturing depends on the manufacturer's ability to provide customers with lean services and lifecycle costs (LCC) for sustainable values at all times. E-maintenance is transforming the maintenance function into a service business to support customers anywhere and anytime with the use of Internet, web-enabled wireless communication and technology. E-maintenance enables the companies with predictive intelligent/embedded sensors to monitor their assets through web-based wireless communication to prevent unexpected breakdown. This system can compare the performance of their product to others, using a global network system to focus on degradation monitoring and prognostics. Such information will greatly facilitate business decisions.

The main problem with performance measurement for decision-making is the non-availability of relevant data and information. However, the recent application of information and communication technology (ICT) and other emerging technologies facilitates easy and effective collection of data and information. The two most important applications of the measurements are the identification of opportunities to improve existing equipment and plants and improved supplier performance. To cite one example, MPM has become part of the decision-making process in the mining industry, where the use of condition monitoring is extremely important.

E-maintenance is a maintenance management concept whereby plants and machineries are monitored and managed by computer software, involving intelligent sensors, databases, wireless communication, Internet, online have produced studies investigating the monitoring of manufacturing performance using wireless communication technologies and the impact on maintenance performance. Today, with the availability of unique e-maintenance solutions, the production and process industry can benefit from server-based software applications, the latest embedded Internet interface devices and state-of-the-art data security. E-maintenance creates a virtual knowledge centre by linking users, technicians/experts and manufacturers. It is useful for the process industry, as it can help to reduce overall costs, ensure savings in resources through OEE and yield a better return on maintenance investment (ROMI).

Some existing e-maintenance solutions provide server based software and equipment embedded internet interface devices (health management card). These e-maintenance solutions provide 24/7 (24 hours a day, 7 days a week) real-time monitoring, controls and alerts. The system converts data into information, available to all concerned for decision making and predicting the performance condition of the plant and machinery on a real time basis. This enables the system to meet supply chain requirements.

Important Maintenance Metrics

Effective maintenance of equipment is a critical factor in delivering quality operations that provide timely resources at a minimal cost. However, those in the maintenance field understand that equipment reliability does not come easy.

Organizations need to set quality benchmarks to measure the current effectiveness and predict future performance and use the data obtained to understand where to make improvements. One way to do this is by using different maintenance metrics to understand the equipment performance. These metrics are very important as they can mean the difference between achieving the overall business goals and explaining how unexpected breakdowns caused yet another production delay.

What are the Maintenance Metrics?

There are two categories of maintenance key performance indicators which include the leading and lagging indicators. The leading indicators signal future events and the lagging indicators follow the past events.

The leading indicator comprises from metrics like the estimated vs actual performance and PM Compliance, while the lagging indicator are reflected in maintenance metrics like the Mean Time To Repair (MTTR), Overall Equipment Effectiveness (OEE) and Mean time between failure (MTBF).

Using these maintenance metrics and turning the data into actionable information, organizations can acquire both qualitative and quantitative insights. And there is no better way to spot opportunities for improvement. Here are some important maintenance metrics you should track if you want to improve and optimize your maintenance operations.

1. Planned maintenance percentage (PPC): This metrics represents the percentage of time spent on planned maintenance activities against the unplanned.

    In simpler terms, this metric tells you how much maintenance work done on a particular asset was a part of your preventive maintenance plan versus how much time you've spent repairing it because it unexpectedly broke down.

    In a great system, 90% of the maintenance should be planned.

    The calculation is as follows:

    PPC= (scheduled maintenance time/total maintenance hours) x 100

2. Overall Equipment Effectiveness (OEE): OEE is the measure of the productivity of a piece of equipment. It gives informed data on how effective organization's maintenance processes is running based on factors like equipment quality, performance, and availability.

A 100% OEE means that your system is producing no defects, as fast as possible, and with no stops in the production. Understanding OEE and the underlying losses, organizations can gain significant insights into how to improve their manufacturing processes. Using this metric, you can identify what has a negative impact on your production, so you can eliminate it.

To calculate the OEE, you multiply the availability by the performance and quality:

OEE = availability x performance x quality

3. Mean time to repair (MTTR): MTTR is the measure of the repairable items maintainability.

   The MTTR clock starts ticking when the repairs start and it goes on until operations are restored. This includes repair time, testing period, and return to the normal operating condition. The goal of every organizations is to reduce MTTR as much as possible. This is especially important for critical assets as ever additional hour you need to restore an asset to a working condition amount to huge losses for your firm.

   To calculate MTTR, you divide the downtime period by the total number of downtimes:

   MTTR= (sum of downtime periods/ total number of repairs)

4. Mean time between failure (MTBF): MTBF is the measure of the predicted time between one breakdown to the next during normal operation.

   In essence, MTBF tells you the expected lifetime for a specific piece of equipment. Higher MTBF means that the part (or product) you bought will work longer before it experiences failure.

   If you know how long a specific part/equipment will last, it gets much easier to predict and prepare for a failure or schedule some preventive work.

   To calculate the MTBF, you divide the total operational time by the number of failures:

   MTBF= (sum of operational time/total number of failures)

5. Preventive maintenance compliance (PMC): PM compliance is defined as the percentage of the preventive work scheduled and completed in a set time.

For example, you might have 60 Work Orders (that are a part of the PM plan) scheduled but 51 completed at the end of the month.

In this case:

PMC= (51/60) x 100 = 85%

This tells you that 85% of all preventive WO's have been covered for selected month.

The disadvantage of this metric is that it doesn't tell you if the WO's have been completed on time.

That is why you need to invest some additional effort and also track if the Work Orders are actually being finished on time. By fat the best way to do that is to use a CMMS as it allows you to quickly create, assign, and track all of your WO's from one place.

Planned Maintenance Percentage

Planned maintenance percentage (PMP) is a maintenance metric that measures the number of planned maintenance tasks in comparison to all maintenance tasks. PMP is expressed as the percentage of total maintenance hours spent on planned maintenance tasks in a given period.

Planned maintenance percentage is a valuable metric for tracking the health of a preventive maintenance program and identifying opportunities to reduce reactive maintenance. PMP can also be used to pinpoint the cause of failure, inefficiencies, and broken maintenance processes so they can be fixed.

How is PMP Calculated?

Planned maintenance percentage is calculated by dividing the total number of planned maintenance hours in a given period by the total number of hours spent on all maintenance in the same period. This number is multiplied by 100 to give you the final percentage.

PMP = # of planned maintenance hours ÷ # of total maintenance hours × 100

$$PMP = \frac{\text{\# of planned maintenance hours}}{} \div \text{\# of total maintenance hours} \times 100$$

For example, if you spent 175 hours during the month on planned maintenance and 200 hours on all maintenance, your planned maintenance percentage would be 87.5%.

Planned maintenance is any maintenance that is organized, documented, and scheduled before asset failure. Any maintenance that is scheduled in response to a breakdown is not considered planned maintenance. Organizations with best-in-class maintenance will have a PMP of 85% or higher. Planned maintenance can also be broken down into several types. Top performing organizations complete about 30% usage-based maintenance, 50% condition-based maintenance, and 5% run to failure maintenance.

How is PMP used?

Planned maintenance percentage is used to make data-driven decisions on maintenance schedules, the allocation of resources, and maintenance processes. Because planned maintenance has a huge impact on the health of assets and how maintenance teams operate, tracking and improving PMP can help organizations maximize asset reliability and control the costs associated with that goal.

What does PMP Mean for Maintenance?

Planned maintenance is a cornerstone of a healthy maintenance operation. It is central to using time efficiently, reducing downtime, staying compliant with safety regulations, and spending money effectively. Measuring planned maintenance percentage allows maintenance teams to gain insight into all these areas so they can continue to improve on their strengths while targeting their weaknesses.

Create Effective Schedules

Schedules can be set up and optimized well in advance when a higher percentage of work is planned. This cuts down on wasted labour and parts while allowing the maintenance team to coordinate its tasks with the production schedule.

Reduce Downtime

Planned maintenance often takes the form of preventive maintenance, which has been shown to cut down on unplanned downtime by up to 18%. Measuring PMP, along with other maintenance metrics, like MTBF and MTTR, allows you to see which assets cause the most downtime and if it could be remedied with more preventive maintenance.

Stay Compliant

Audits come in many shapes and sizes, but they all rely on two basic components: real-time observation and historical proof. Using planned maintenance percentage is a key way the maintenance team can ensure assets are routinely checked and that there is a comprehensive log that shows these efforts. Both are integral for passing audits and staying compliant.

Control Costs

PMP can be used to control costs in two ways—resource efficiency and downtime reduction. A higher PMP means an increase in planned tasks, and the ability to schedule labour and parts far into the future. This helps to reduce instances of over-scheduling and over-purchasing, saving you money. Reducing downtime with planned maintenance leads to more uptime, increased production, and less money lost from stoppages.

Bottom Line on Planned Maintenance Percentage

Measuring, analyzing, and improving planned maintenance percentage can have a huge domino effect on your entire facility. Moving away from reactive maintenance to a proactive approach gives your operation more control over its tasks, resources, and money. It also gives critical equipment the attention it deserves and benefits every part of the business, from production to customer service, finance, and beyond.

Preventive Maintenance Compliance

Preventive maintenance compliance (PMC) is the percentage of preventive maintenance work orders that were executed within a schedule. We will explain how to calculate your preventive maintenance compliance score, which rules you should follow and why PMC is so important to keep your company firing on all cylinders.

Calculate the Preventive Maintenance Compliance Score

The 4 steps to calculate the preventive maintenance compliance score include:

1. Defining a preventive maintenance plan and a list of tasks to be performed on each asset.

2. Executing the preventive maintenance plan in your previously established time period.

3. Monitoring the number of tasks completed on time.

4. Calculating the percentage of maintenance work orders completed within schedule.

It's not enough to just know whether the preventive maintenance tasks were completed or not. You also need to know which ones were actually completed on time and when. You should try to ensure that all preventive maintenance work is done ahead of schedule. As a rule of thumb, you should try ensure that the time it takes to complete a equates to 10% of the total time, i.e., if you're given 100 days to complete a maintenance task, try to do it at least 10 days ahead of schedule.

Why is this Golden Rule so Important in Preventive Maintenance?

The 10% rule didn't come out of nowhere. We know that even at peak performance only 90% of maintenance tasks are preventive. The remaining 10% is corrective maintenance due to failures that can't be predicted. So these guidlines ensure that your team has room for manoeuver in the event of any unexpected failures. You should aim for an 80/20 ratio — 80% preventive maintenance, 20% corrective maintenance.

You should make sure that your MTTR (mean time to repair) and MTBF (mean time between failures) are both healthy. However, it's best not to take risks. If you don't comply with the preventive maintenance plan, you risk accumulating dozens of problems and irreparably harming the operation of the company.

Implementation of Effective Maintenance Planning and Scheduling

Many organizations have implemented planning and scheduling into their maintenance operations to improve the utilization of their personnel, reduce unplanned downtime and more effectively manage their maintenance budgets. Every company experiences obstacles, roadblocks and resistance along the way, and some are more successful than others.

The concept of planning and scheduling maintenance activities is not a novel idea. In fact, this process has been around for years. In some form, all maintenance work is planned and scheduled. But who does the planning and scheduling, when does it take place and is the resultant work executed in the most effective, efficient and cost-conscious way? The company, which has multiple factories in the upper Midwest, realized that it needed to address the way it managed its maintenance activities in order to remain competitive in a global marketplace. Work was not tracked accurately, spare parts were not associated with equipment, labor and material costs were not tracked to the assets, and budgets were overrun on most outages.

The company's initial approach was to conduct an assessment of its current maintenance management processes, identify the major gaps and develop an action plan to tackle those gaps. The assessment took place in 1999 and ranked the company 437 out of 1,000, which was just out of the "reactive" scale. The corporate maintenance manager took the assessment information and recommendations into account to formulate his action plan.

Challenges Encountered

Buy-in to implement the maintenance manager's plan was not easy to obtain, both at the corporate and factory management level. There was resistance in changing the way maintenance had been managed for many years. Another challenge involved getting all the factories aligned. Every factory had its own ways of performing maintenance activities, and in order to measure the progress of this initiative, each would be required to follow the same template. These challenges all warranted training on what best practice should look like.

Approach

The maintenance manager arranged a visit to an aluminum smelter in South Carolina that had been identified as one of the nation's best. Delegates from the company made the visit and observed the smelter's maintenance practices as well as the roles, responsibilities and duties of its planners. After seeing how the planner/scheduler worked at the smelter, the company created a planner position at each of its factories. The planners were selected from the maintenance workforce by seniority. It became evident early on that one planner was not adequate to support the maintenance efforts of the factories. Additional planner positions were posted at each location. Rather than basing the selection of individuals solely on seniority, additional criteria were established to obtain the best individuals for the role.

References

- Planning-and-scheduling, implementation, consulting-implementation-training: idcon.com, Retrieved 25 February, 2019

- 5-tips-for-more-effective-maintenance-planning-and-scheduling, maintenance-management: assetivity.com.au, Retrieved 29 March, 2019

- Equipment-maintenance-strategy, maintenance-management, free-articles: lifetime-reliability.com, Retrieved 13 February, 2019

- Effective-maintenance-plan, read: reliableplant.com, Retrieved 17 May, 2019

- Set-up-preventive-maintenance-plan: micromain.com, Retrieved 13 July, 2019

- The-importance-of-planned-maintenance: baass.com, Retrieved 3 June, 2019

- Top-10-signs-your-maintenance-planning-and-scheduling-isnt-working, posts: prometheus-group.com, Retrieved 16 January, 2019

- 6-maintenance-scheduling-principles-to-improve-overall-workforce-efficiency, posts: prometheus-group.com, Retrieved 19 May, 2019

- 5-important-maintenance-metrics-and-how-to-use-them, applications: maintworld.com, Retrieved 28 April, 2019

4

Maintenance Optimization

The analysis and development of strategic models for the improvement of maintenance policies is known as maintenance optimization. It can be categorized into planned maintenance optimization and preventive maintenance optimization. This chapter closely examines maintenance optimization to provide an extensive understanding of the subject.

Decision models can help companies determine the value of maintenance. There are many decision models designed to determine optimium maintenance policies in the literature. Maintenance optimization models are mathematical models which quanitfies both the drawbacks (costs) and benefits of maintenance, and allows an optimum balance between the two to be found. Indeed, Dekker defines maintenance optimizati.on models as follows:

> "Maintenance optimization models are mathemetical models whose aim it is to find the optimum balance beween costs and benefits of maintenance, while taking all sorts of constraints into account".

However, the gap between academic models and application in a business specific context is currently the biggest problem encountered in the field of maintenance optimization. One reason for this gap is that many papers have been written for maths purposes only. Mathematical analysis and techniques, rather than solutions to real problems, have been the focus of many papers on maintenance optimization models. It can be argued that maintenance optimization should not start with developing a maintenance model and trying to fit an application to it, but it should start with an application and try to fit a maintenance optimization model to it.

Maintenance optimization consists of more than a maintenance optimization model. The main ingredients of a maintenance optimization method are the chosen performance indicators, their predictive models, the selected types of maintenance, the optimization problem formulation, and the solution technique.

Optimization Criteria

Maintenance objectives can be divided into five categories ensuring system function (availability, efficiency and product quality), ensuring system life (asset management), ensuring safety, ensuring human well-being and ensuring optimal capital replacement decisions.

The maintenance objectives are quantified by corresponding maintenance performance indicators, or a combination of maintenance performance indicators. Maintenance performance indicators should ideally be selected to represent only the performance of the maintenance organization, and be independent of non-related effects such as differences in production volumes or changes to the operating environment. This is a tricky task, as the benefits of maintenance improvement tend to show up in other areas like production, quality or inventory. The maintenance organization itself may show, for example, higher cost. Another problem is knowing which maintenance criteria are the most relevant for a particular maintenance organization. In an attempt to alleviate these problems Horenbeek, Liliane Pintelon, and Muchiri has created a generic list of maintenance optimization criteria to serve as a starting point when selecting objectives for a particular problem:

- Maintenance costs (discounted),
- Maintenance quality,
- Personnel management,
- Inventory of spare parts,
- Overall equipment effectiveness,
- Number of maintenance interventions,
- Capital replacement decisions,
- Asset management,
- Availability,
- Reliability,
- Maintainability,
- Environmental impact,
- Safety/risk,
- Logistics,
- Output quantity,
- Output quality.

A large number of maintenance optimization models have been published over the years, but the vast majority of them focus on a single optimization objective. Multi-objective optimization models, in particular those considering multiple-components, is an underexplored area of maintenance optimization.

Multi-component Maintenance Models

While there are plenty of maintenance optimization models focusing on single components, typically to determine optimal replacement intervals of that component, any real world maintenance management situation requires a focus on multiple components and multiple systems. With increasing complexity of the maintained equipment, maintenance optimization models need to be able to describe the interaction between multiple components and units. Cho and Parlar defined multi-component maintenance optimization models as follows:

> "Multi-component maintenance models are concerned with optimal maintenance policies for a system consisting of several units of machines or many pieces of equipment, which may or may not depend on each other (economically/stochastically/structurally)."

The fact that up to 10 years ago, the vast majority of the maintenance models were concerned with one single piece of equipment operating in a fixed environment was considered as an intrinsic barrier for applications. However, currently, there seem to be very few models that consider multiple types of dependencies within the same model. There are some recent works, but it is a relatively unexplored area of maintenance modeling.

Economic Dependence

Economic dependence implies that costs can be saved when several components are jointly maintained instead of separately. Alternatively, the opposite can be true so that simultaneous downtime of components is undesirable and hence maintenance must be spread out over time as much as possible. In the first case, economic dependence means that benefits can be achieved by grouping maintenance actions together. In the second, grouping maintenance likely leads to higher costs.

Stochastic Dependence

Stochastic dependence occurs if the condition of some components influences the lifetime distribution of other components. Stochastic dependence is sometimes referred to as failure interaction or probabilistic dependence. The degradation or failure of one component will in turn cause failure, degradation or an increased wear rate of stochastically dependent components.

Structural Dependence

Structural dependence applies if components structurally form a part, so that maintenance of a failed component implies maintenance of working components. Structural dependence thus interferes with a maintenance managers options when it comes to grouping of maintenance actions that may be desirable from an economic dependence point of view. One example of structural dependence is replacement units, where multiple parts are treated as one at the time of replacement.

Maintenance Decision Support Systems

Decision support systems can help an individual make better decisions through enhanced situational awareness; problem recognition, problem structure, information management, statistical tools, and by suggesting solutions through application of knowledge and optimization techniques. However, the nature of building a decision support system for a specific decision process, and tailoring it to specific managers, makes it hard to identify a generalized approach to building decision support systems. To help structure the area, decision support systems can be classified into passive, active or cooperative decision support systems. A passive decision support system is a system that aids the process of decision making, but cannot bring out explicit suggestions or solutions. An active decision support system has this ability, while a cooperative decision support system allows the decision maker to modify and refine suggestions provided by the system.

Passive Decision Support Systems

Passive decision support systems assists the decision maker, for example by collating and filtering relevant information, but does not explicitly provide suggested improvements or solutions. An abundance of examples of passive decision support systems can be found among the large number of computerized maintenance management systems, devoted to help managing maintenance activities. However, maintenance optimization is typically not included as a feature in these software packages, which makes them excellent databases to track repair orders and maintaining appropriate book-keeping.

Active Decision Support Systems

Active decision support systems explicitly suggest improved solutions. If a maintenance optimization using a single-objective model is performed, a single best solution would be presented as the outcome. However, computers cannot think about reality. They work with a model and human operators are required to relate the model to reality.

A major shortcoming of most maintenance decision support systems is that they act like a black box. Since each maintenance problem is likely to be different, it is only the user (and not the decision support system) who can validate the calculations and convince his/her management of their value. Similar concerns have been raised among the Swedish armed forces, where several decision support systems have have already been developed and marketed but not always with success. Interviews performed with Swedish officers suggested one reason for these failures; human contributions cannot be excluded from the analysis and are often vital for the result. An officer expressed his experience succinctly: "A decision support system should give *me* time to think".

Cooperative Decision Support Systems

A cooperative decision support system allows the decision maker to modify and refine suggestions provided by the system. This helps alleviating the problems users experience in applying the results of active decision support systems to the real world situation, by facilitating integration of the human contributions into the decision making process.

By utilizing a multi-criteria maintenance optimization model, and a corresponding multi-criteria optimization algorithm, a Pareto-optimal set of suggested solutions can be generated. The benefits of simultaneous optimization of different criteria has been noted in the literature. For example, Wang notes that to achieve the best operating performance, an optimal maintenance policy needs to consider both maintenance cost and reliability measures simultaneously, and D. Murthy, Atrens, and Eccleston argue that operating load and maintenance strategies need to be optimized jointly since the load degrades the equipment and maintenance actions control the degradation. By creating and presenting a Pareto-optimal set of solutions, the user will be able to perform a trade-off analysis among the modeled criteria, and selecting candidate solutions for implementation by integrating real world knowledge not included in the model.

Decision models can help companies determine the value of maintenance. Maintenance optimization models are mathemetical models whose aim it is to find the optimum balance beween costs and benefits of maintenance, while taking all sorts of constraints into account. The main ingredients of a maintenance optimization method are the chosen performance indicators, their predictive models, the selected types of maintenance, the optimization problem formulation, and the solution technique.

Multi-objective optimization models, in particular those considering multiple components, is an underexplored area of maintenance optimization. Multi-component maintenance models are concerned with optimal maintenance policies for a system consisting of several units of machines or many pieces of equipment, which may or may not depend on each other (economically/stochastically/structurally). Decision support systems can be classified into passive, active or cooperative decision support systems. A passive decision support system is a system that aids the process of decision making, but cannot bring out explicit suggestions or solutions. An active decision support system has this ability, while a cooperative decision support system allows the decision maker to modify and refine suggestions provided by the system. Creation of a cooperative decision support system will be facilitated by the application of multi-objective optimization to create a Pareto-optimal set of solutions.

Maintenance Optimization Model

Unlike the situation that exists with Reliability Centred Maintenance, there are no internationally recognised standards for performing Preventive Maintenance Optimisation. A quick literature search using the term Preventive Maintenance Optimisation reveals that there are a number of scholarly informations on the topic, and a wide range of approaches taken by different vendors and consultants in the field.

These approaches can be divided into two camps:

- Quantitative or Stochastic Approaches – Which typically use statistics, hazard functions, Weibull Analysis, Monte Carlo analysis and Markov models to simulate failure probabilities and consequences over time.

- Qualitative (or semi-quantitative) Approaches – Which often apply key Reliability Centred Maintenance concepts and approaches, but which seek to arrive at the results more quickly than would be the case if using a full RCM approach.

Our only comments regarding quantitative approaches is to note that these can only be effective if there is reasonable data – and in particular, reasonable data regarding equipment failures or impending failures. While this may be possible where large fleets of essentially identical equipment operate under essentially identical operating conditions, the situations where this is true is comparatively few and far between. Large fleets of mobile equipment (such as trucks, or railway rolling stock) may be possible candidates for this type of analysis, but once again, depend heavily on the existence of complete and reliable data. Assetivity's preferred approach to PM Optimisation is to conduct a type of "reverse RCM".

This approach consists of asking the following seven questions regarding the asset or system under review:

- Current PM Tasks - What are the current PM tasks being performed?

- Failure Modes - What failure modes are these addressing?

- In-service Failures – What in-service failures are currently being experienced, and what are their causes?

- Hidden Failures - What protective devices and systems are in place and what are the potential failure modes associated with these?

- Failure Consequences - In what way does each failure matter?

- Proactive Tasks - What can be done to predict or prevent each failure?

- Default Actions - What should be done if a suitable proactive task cannot be found?

Current PM Tasks

One of the key concepts of Reliability Centred Maintenance is that all preventive maintenance tasks should be directed at preventing specific failure modes (or causes of failure). In PM Optimisation, as the primary objective is to review and optimise the current Preventive Maintenance program, the starting point is to begin by assembling all of the tasks that this Preventive Maintenance program consists of. Note that it will be essential to be quite specific regarding the tasks being undertaken. For example, a

Preventive Maintenance task described as "500-hour service" would need to be broken down into the various tasks that the 500-hour service consists of. This needs to be done so that (in the next step) the Failure Mode (or Modes) that each task is addressing can be identified and critically reviewed.

These tasks (and the detail contained within them) may exist in a number of disparate systems. Ideally, they would all be contained within your CMMS or EAM system, but often they are not. For example, the CMMS may contain a PM activity titled "Perform Monthly Vibration Analysis", but the details of what equipment is analysed, and what it is analysed for may be contained in a separate Condition Monitoring System. It is also important to include routine maintenance or equipment inspection activities performed by plant operators in this analysis, and the details of these are usually not contained within your CMMS or EAM system.

Failure Modes

Once each current PM tasks has been identified, the next step in the PM Optimisation process is to identify the Failure Modes which these tasks are intended to address. In this context, note that we consider a "Failure Mode" to be a cause of Failure. In performing this analysis it is important to limit the analysis to failure modes which are "reasonably likely" to occur on the same or similar equipment operating in the same context. This requires elements of experience and engineering judgement. It is also important to ensure that causes are described in enough detail to ensure that time and effort are not wasted trying to treat symptoms instead of causes.

In-service Failures

In this step, we need to add to the list of failure modes under consideration those that have not been prevented as a result of the Preventive Maintenance routines that have been put in place. This requires us to obtain details of the Failure History for the equipment or system being analysed. More specifically, it also requires us to be able to identify the causes of those failures. The starting point for this is to obtain from your CMMS or EAM details of the Breakdown and Unplanned Work Orders that have been raised against the equipment. However, we should also recognise that, in the real world, in most industries, this Work Order history is likely to be incomplete, possibly inaccurate, and is also unlikely to accurately record the cause of equipment failure (at least to the level of detail that we require when reviewing the PM program). It is highly likely, at this point, that you will need to supplement the data contained in your corporate information systems with the knowledge held in the heads of those that operate and maintain the plant. Another potentially useful source of information regarding equipment failures is your organisation's Downtime or Loss Accounting log, if it has one. However, once again, it is highly advisable to combine this with the knowledge and experience of your plant operators and maintainers in order to ensure that you get a complete and accurate picture of what failures are occurring on the equipment.

For each failure, you should identify the Failure Mode (or cause) and add it to the list generated in Step 2 of this process.

Hidden Failures

There is one final class of failure modes that you should consider in your PM Optimisation analysis. These are failure modes that may not be being prevented through existing PM (so won't appears in the list of failure modes generated in Step 2), and may have actually happened, but won't appear in your failure history (and so won't be added in Step 3). They are Hidden Failure modes. In RCM terminology, a Hidden Failure is one which, when it happens on its own, will not become apparent under normal circumstances (unless a PM task is in place to detect it). An example of this is failure of a smoke detector. If no PM task was in place to periodically test the smoke detector, then its failure (so that it could not detect a fire even if one were to occur) would not become apparent until a fire occurred (and the consequences of this could, potentially, be disastrous). Hidden Failures are most often associated with Protective Devices and Systems which are not fail-safe. This could include alarms, interlocks, trips, stand-by equipment and many other devices and systems.

To ensure that we consider Hidden Failures in our PM Optimisation analysis, we strongly recommend that time be taken to specifically seek out protective devices and systems on the equipment being analysed, and identify failure modes that are associated with these systems that are Hidden. These failure modes should be added to the list of Failure Modes generated in Steps 2 and 3.

So far in this process, we have spent time identifying the Failure Modes for which the Preventive Maintenance program needs to be optimised. It has been, up to this point, largely a data collection exercise. In the following steps, we now start to analyse this data and make decisions regarding the most appropriate PM tasks to put in place to address each Failure Mode. The observant amongst you will recognise that the following three steps are essentially identical to the final three steps in the RCM process.

Failure Consequences

Another key concept underpinning Reliability Centred Maintenance is that the primary objective of a Preventive Maintenance program is not necessarily to avoid or minimise failures themselves, but to avoid or minimise the consequences of those failures. There is little point in spending a lot of time and money preventing failures that have little or no consequences associated with them. On the other hand, if a failure has serious consequences, we may be able to justify going to great lengths to avoid those consequences. In this way, the RCM process focuses attention on the maintenance activities which have most effect on the performance of the organization, and diverts energy away from those which have little or no effect.

This fifth step in the RCM process classifies the consequences associated with each failure mode as belonging to one of the following four groups:

- Hidden failure consequences: Hidden failures have no direct impact, but they expose the organization to multiple failures with serious, often catastrophic, consequences. (Most of these failures are associated with protective devices which are not fail-safe).

- Safety and environmental consequences: A failure has safety consequences if it could hurt or kill someone. It has environmental consequences if it could lead to a breach of any corporate, regional, national or international environmental standard.

- Operational consequences: A failure has operational consequences if it affects production (output, product quality, customer service or operating costs in addition to the direct cost of repair).

- Non-operational consequences: Evident failures which fall into this category affect neither safety nor production, so they involve only the direct cost of repair.

In making these decisions, we need to consider what the potential effects would be for each failure mode. This involves considering:

- What evidence is there (if any) that the failure has occurred?

- In what ways (if any) does the failure mode poses a threat to safety or the environment?

- In what ways (if any) does the failure mode affects production or operations?

- What physical damage (if any) is caused by the failure?

- What must be done to repair the failure?

The consequence evaluation process shifts emphasis away from the idea that all failures are bad and must be prevented. In so doing, it focuses attention on the maintenance activities which have most effect on the performance of the organization, and diverts energy away from those which have little or no effect.

Proactive Tasks

In PM Optimisation (and RCM), failure management techniques are divided into two categories:

- Proactive Tasks: These are tasks undertaken before a failure occurs, in order to prevent the item from getting into a failed state. They embrace what is traditionally known as 'predictive' and 'preventive' maintenance, although we will see later that RCM uses the terms Scheduled Restoration, Scheduled Discard, and Condition-based Maintenance.

- Default Actions: These deal with the failed state, and are chosen when it is not possible to identify an effective proactive task. Default actions include failure-finding, redesign and run-to-failure.

Many people still believe that the best way to improve equipment reliability is to do some kind of proactive maintenance on a routine basis. Conventional wisdom suggested that this should consist of overhauls or component replacements at fixed intervals. Figure illustrates the fixed interval view of failure.

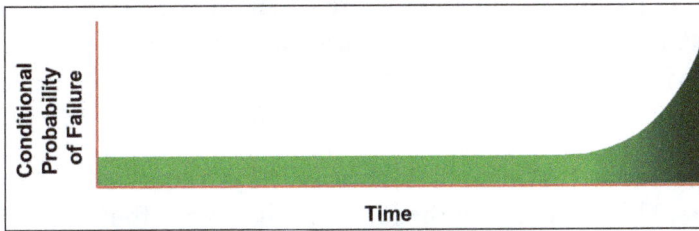

Traditional View of Equipment Failure.

Figure is based on the assumption that most items operate reliably for a period of time, and then wear out. Classical thinking suggests that extensive records about failure will enable us to determine this life and so make plans to take preventive action shortly before the item is due to fail in future.

This model is true for certain types of simple equipment, and for some complex items with dominant failure modes. In particular, wear-out characteristics are often found where equipment comes into direct contact with the product. Age-related failures are also often associated with fatigue, corrosion, abrasion and evaporation.

However, equipment in general is far more complex than it used to be. This has led to significant changes in the patterns of failure, as shown in figure. The graphs show conditional probability of failure against operating age for a variety of electrical and mechanical items.

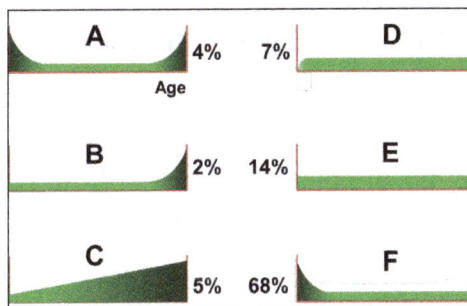

Six Failure Patterns of RCM.

Pattern A is the well-known bathtub curve. It begins with a high incidence of failure (known as infant mortality) followed by a constant or gradually increasing conditional probability of failure, then by a wear-out zone. Pattern B shows constant or slowly increasing conditional probability of failure, ending in a wear-out zone.

Pattern C shows slowly increasing conditional probability of failure, but there is no identifiable wear-out age. Pattern D shows low conditional probability of failure when the item is new or has just been refurbished, then a rapid increase to a constant level, while pattern E shows a constant conditional probability of failure at all ages (random failure). Pattern F starts with high infant mortality, which drops eventually to a constant or very slowly increasing conditional probability of failure.

Studies done by Nowlan and Heap in the 1960s on civil aircraft showed that 4% of the items conformed to pattern A, 2% to B, 5% to C, 7% to D, 14% to E and no fewer than 68% to pattern F. (The number of times these patterns occur in aircraft is not necessarily the same as in other industries. But there is no doubt that as assets become more complex, we see more and more of patterns E and F).

These findings contradict the belief that there is always a connection between reliability and operating age. This belief led to the idea that the more often an item is overhauled, the less likely it is to fail. Nowadays, this is seldom true. Unless there is a dominant age-related failure mode, age limits do little or nothing to improve the reliability of complex items. In fact, scheduled overhauls can actually increase overall failure rates by introducing infant mortality into otherwise stable systems.

PM Optimisation and RCM divide proactive tasks into three categories, as follows:

- Scheduled Restoration tasks.

- Scheduled Discard tasks.

- Condition-based Maintenance tasks.

- Scheduled Restoration and Scheduled Discard tasks.

Scheduled restoration entails remanufacturing a component or overhauling an assembly at or before a specified age limit, regardless of its condition at the time. Similarly, scheduled discard entails discarding an item at or before a specified life limit, regardless of its condition at the time.

Collectively, these two types of tasks are now generally known as Preventive Maintenance. They used to be by far the most widely used form of proactive maintenance. However, they are much less widely used than they used to be.

The continuing need to prevent certain types of failure, and the growing inability of classical techniques to do so, are behind the growth of Condition-based approaches to failure management. The majority of these techniques rely on the fact that most failures give some warning of the fact that they are about to occur. These warnings are known as Potential Failures, and are defined as identifiable physical conditions which indicate that a Functional Failure is about to occur or is in the process of occurring.

The new techniques are used to detect potential failures so that action can be taken to avoid the consequences which could occur if they degenerate into Functional Failures. They are called Condition-based tasks because items are left in service on the condition that they continue to meet desired performance standards. Condition-based tasks can include the use of sophisticated technology, such as: Vibration Analysis, Thermography, Oil Analysis, Ultrasonics and others, but they can also include simple techniques such as visual inspection. Used appropriately, on-condition tasks are a very good way of managing failures, but they can also, if not applied in the right way and at the right frequency, be an expensive waste of time.

In PM Optimisation, we use the same structured decision-making process that is used in RCM. This has clear evaluation criteria and enables decisions regarding the selection of the appropriate Proactive Task to be made with confidence.

Planned Maintenance Optimization

Planned Maintenance Optimization (PMO) is a method of improving maintenance strategies based on existing preventive maintenance (PM) routines and available failure history. While most companies have identified the need for a preventive maintenance (PM) program, the effective execution of such maintenance activities can be challenging given the everyday demands of a facility. Unseen circumstances that require urgent attention can easily derail planned activities and can potentially disrupt a smoothly running plant.

While alternatives such as reliability centered maintenance (RCM) addresses some of the factors that make PM a cost- and labor-intensive process, coming up with a robust RCM strategy may take long periods of time.

PMO provides a method through which maintenance activities are carried out more efficiently. By performing PMO, a new maintenance strategy is derived from existing PM tasks. Given the existing tasks, modifications on the schedule and frequency of the routines are done based on the failure history of the equipment. With a relatively shorter time to develop, the resulting strategy can be similar to performing RCM.

Three Phases of PMO

The PMO process can be summarized in three phases:

1. Data collection: Any attempt at optimization starts with good, reliable data. Data on equipment performance, particularly on failure history over time, must be collected. A minimum time period must be set to ensure that enough insight is obtained from the data. Tools such as a CMMS program can make this process easier and more accurate.

2. Data analysis, review and recommendations: The collected data must be analyzed to identify which equipment is the most critical. Some points to consider are criticality to the plant's operations, cost to repair, MTBF, and MTR.

    The information gathered from analyzing the data must then be reviewed against existing PM routines. Some key points to review are: a) whether the PM routines are scheduled correctly to align with the MTBF and MTR data points, and b) whether failure points are within acceptable tolerances set by original equipment manufacturer (OEM) specifications or industry standards. Any substantial deviations from such checks can be a source of improvement from a maintenance standpoint. Based on the review, recommendations on modifications for the PM tasks should be made.

    Schedules and frequencies of activities need to be optimized to meet MTBF and MTR constraints. Any missing maintenance activities, as well as redundancies in tasks, need to be addressed accordingly.

3. Agreement and execution: Agreed action items must be delegated properly. Identified task owners should be accountable for any required action and monitored for progress. Note that the PMO process is a continuous effort and reviews should be done habitually.

Benefits of Applying PMO

Regular maintenance activities are clearly a key part in ensuring a plant's reliability. But PMO further increases the benefits of maintenance activities by showing substantial reductions in costs. In the laboratory and life sciences industry, a PMO program is estimated to reduce overall maintenance costs by around 25%. Payback periods of investing in a PMO strategy are estimated at around 12 to 24 months, just considering the measured savings from maintenance costs.

Aside from the improvements in uptime and reliability that come with a robust maintenance strategy, PMO methods enable company resources to be spent more wisely without sacrificing the quality of execution of maintenance tasks.

Maintenance activities, particularly PM activities, are already proven concepts that increase the overall performance of a plant. With continuous practice, PMO is a tool that can help execute PM activities more efficiently and effectively.

Preventive Maintenance Optimization

In order to balance the risk of failure and the costs of maintenance, programs should be periodically reviewed based on failure history and performance analysis. This is called Preventive Maintenance Optimization. Preventive Maintenance Optimization (PMO) is a method of continuous improvement, working to increase the effectiveness and efficiency of maintenance activities. In addition, PMO:

- Increases cost effectiveness.
- Improves reliability.
- Increases machine uptime.
- Enhances an organization's understanding of the level of risk they are managing.
- Reduces the ambiguity of maintenance tasks that are not clearly written.
- Helps avoid or eliminate redundant PM and condition based maintenance tasks.
- Allows for the refocusing of resources toward failure prevention maintenance activities.

Approaches to Optimizing PM Schedules

There are three popular approaches for optimizing Preventive Maintenance schedules:

- Reliability-centered Maintenance (RCM).
- Failure reporting and corrective action system (FRACAS).
- Judgment-based approach.

Reliability-centered Maintenance (RCM)

The reliability-centered maintenance (RCM) approach works to ensure systems continue to do what is required for operations. The goal of developing an RCM program is to implement a unique maintenance schedule for each critical asset within a facility or organization. In his book RCM2, the late John Moubray characterized reliability-centered maintenance as a process to establish the safe minimum levels of maintenance.

A program must meet these four basic principles in order to be recognized as true RCM:

- The program is scoped and structured to preserve system function.
- It identifies how functions are defeated (failure modes).
- It addresses failure modes by importance.
- For important failure modes, it defines applicable maintenance task candidates and selects the most effective one.

A guide to the Reliability-centered Maintenance (RCM) Standard identified the basic requirements a program must meet before it is truly RCM. It begins with these seven questions:

- What is the item supposed to do and its associated performance standards?

- In what ways can it fail to provide the required functions?

- What are the events that cause each failure?

- What happens when each failure occurs?

- In what way does each failure matter?

- What systematic tasks can be performed proactively to prevent, or to diminish to a satisfactory degree, the consequences of the failure?

- What must be done if a suitable preventive task cannot be found?

There are four phases of an RCM project:

- Decision: Justification and planning based on need, readiness and desired outcomes.

- Analysis: Conduct the RCM study in a way that provides a high quality output.

- Implementation: Act on the study's recommendations to update asset and maintenance systems, procedures and design improvements.

- Benefits: Measure the improvements and identify opportunities to further improve.

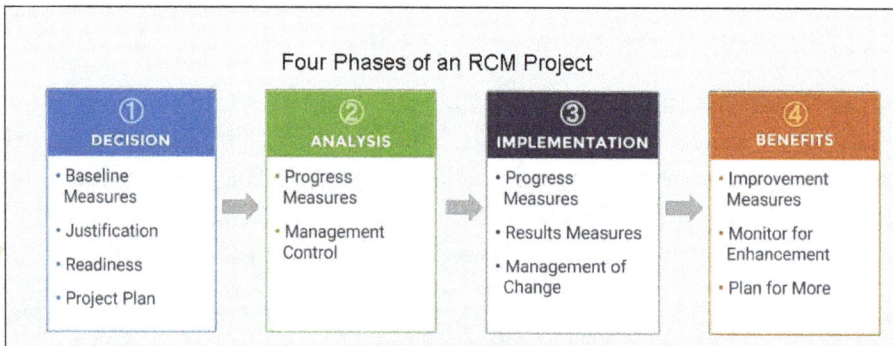

Four Phases of an RCM Project

① DECISION	② ANALYSIS	③ IMPLEMENTATION	④ BENEFITS
• Baseline Measures	• Progress Measures	• Progress Measures	• Improvement Measures
• Justification	• Management Control	• Results Measures	• Monitor for Enhancement
• Readiness		• Management of Change	• Plan for More
• Project Plan			

The RCM approach is a multi-faceted process that requires time, effort and buy in from corporate, executive stakeholders, your maintenance team and more. Because of that, RCM is best utilized to improve efficiency for large scale, capital projects or for critical equipment.

Failure Reporting and Corrective Action System

A failure reporting, analysis and corrective action system (FRACAS) is a more rapid approach to optimize PMs, and does not require extensive planning and decision-making to assess a PM program. FRACAS is a system that establishes a procedure for reporting, classifying, analyzing failures, and planning corrective actions in response to common failures. It identifies the root causes and failure analyses to help organizations implement the best solution to prevent or predict the issue from occurring time and time again.

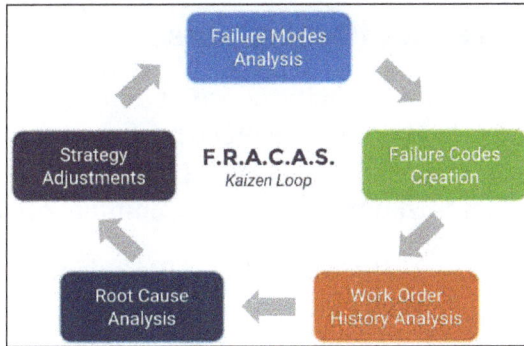

FRACAS consists of:

- Failure Reporting (FR): Asset or system failures are formally reported through a Defect Report, Failure Report or within a Computerized Maintenance Management System (CMMS).

- Analysis (A): Perform analysis in order to identify the root cause of failure.

- Corrective Actions (CA): Identify, implement and verify corrective actions to prevent more repetition of the same failure.

Common outputs from FRACAS include important key performance metrics such as mean time between failures, mean time between repair, mean time to repair, reliability growth, and failure/incidents distribution by type. The FRACAS model provides the information needed to support Root Cause Failure Analysis (RCFA) and Reliability Centered Maintenance (RCM) efforts.

Judgment-based Approach

A judgement-based approach to PM optimization is just as it sounds. The process involves consulting the maintenance team and system engineers to develop a plan based on how they have seen the equipment operate and respond to the current PM schedule.

The judgement approach is not as accurate as RCM or FRACAS, and will not produce the same data and indicators of performance. However, sometimes that data and spending the time to track it is not necessary for all pieces of equipment. Judgement based PM optimization is best for assets that lack criticality, are inexpensive to repair/replace, or have been operating normally.

How a CMMS can Help

Once you have completed the six steps to design a preventive maintenance schedule for your organization, it is important to consistently schedule, track and analyze this information. The functionality within a CMMS can help facilitate this process in a variety of ways.

Automate PM Processes and Procedures

A CMMS offers automation tools to help reduce missing scheduled work and equipment failures, making PM optimization as efficient and streamlined as possible. PM Task Generation, PM Scheduling and Inspections help facilitate continuous improvement and support for an organization's Preventive Maintenance program:

- PM Task Generation – Users can utilize a PM calendar- and meter-based PM tasks for all assets. These include detailed descriptions with how-to's, guidelines and other information vital to effectively performing the work.

- PM Task Schedules – PM schedules empower users to coordinate labor resources and parts needed to complete work, as well as automatically generate PM tasks based on usage or a daily, weekly or monthly basis.

- Inspections – Users can record inspections accurately and generate corrective work orders when equipment inspections fail.

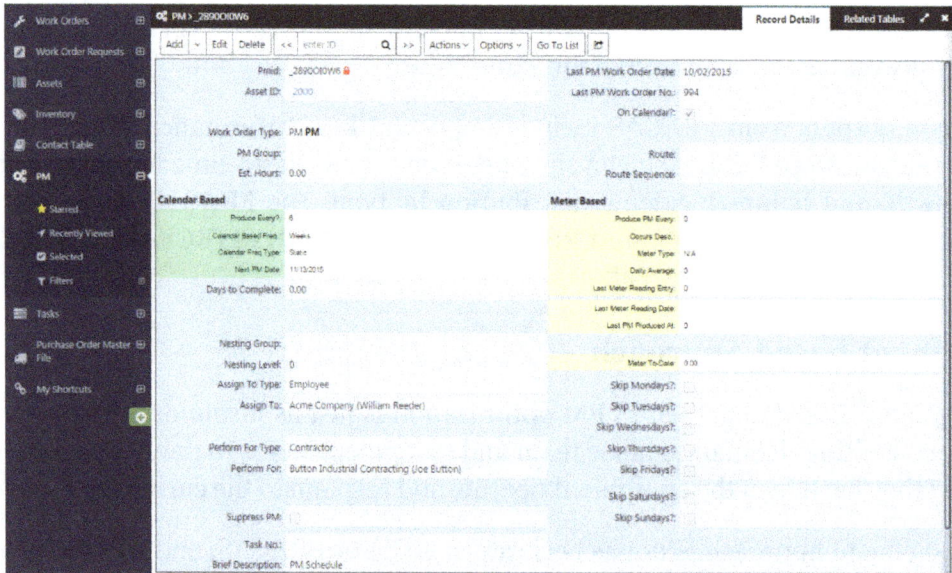

A CMMS completely automates the Failure Reporting step in the FRACAS PM optimization method. Organizations can also leverage a CMMS to perform root cause analysis and deploy corrective actions.

Track and Analyze PM Schedules

Both RCM and FRACAS require considerable tracking and analyzing of PM processes. With eMaint's reporting and dashboard tools, organizations can consistently document work order history, failures, costs and trends. With a few clicks of a mouse, organizations have access the data to perform the analyses that both RCM and FRACAS requires.

- Reports – eMaint's reporting tools help organizations build reports that can be automatically updated with your organization's most important Key Performance Indicators (KPIs) and keep all critical data in one place. You can also auto-generate and email reports to anyone in your organization.

- Dashboards – Dashboards and associated graphs can be generated with your CMMS data on total downtime by critical asset, Mean Time Between Failures (MTBF), wrench time percentage and more. This will help your organization get a better understanding on how equipment is performing and what decisions need to be made to improve processes.

Putting it simply, PMO involves performing the right work, at the right frequency, the right way. Regardless of the approach a company decides upon, optimizing Preventive Maintenance schedules will benefit your organization. For example, eMaint clients experience KPI improvements from PM optimizations such as:

- 30% reduction in corrective maintenance.

- 20% to 80% increase in planned maintenance percentage.

- 50% increase in Return on Net Assets.

- 100% compliance for SLAs.

- 77% increase in PM compliance.

Reliability-based Optimization of Maintenance Scheduling

The application of alternating loading to metallic components may lead to fatigue failure. One or several fatigue cracks initiate and grow within the structure, and finally lead to loss of serviceability or eventually to structural collapse. The occurrence of initial crack, the initiation and propagation of fatigue cracks is a highly uncertain phenomenon and thus, must be addressed within an appropriate concept that accounts for this uncertainty. In particular, the effects of uncertainty can be quantified in terms of structural reliability. As cracks develop and grow during the life time of a structure, a time variant decay of the reliability is to be expected. The harmful effects of propagating cracks can be avoided by scheduling maintenance activities. The scheduling of these activities involves selecting a crack detection technique (e.g. visual inspection, ultrasonic methods, etc.) and an inspection periodicity (e.g. monthly, annual inspection, etc). Among different inspection approaches, Non Destructive Inspection (NDI) techniques play a fundamental role. However, these techniques can fail in detecting cracks. Thus, they are characterized by the probability of detection, which depends on the crack length.

Maintenance activities are necessary to ensure sufficient reliability. However, such activities contribute significantly to the costs associated with the operation of the structure. The best maintenance schedule can be interpreted as a tradeoff between the costs related to the inspection and repair activities and the level of reliability. The high level of uncertainties inherent in the fatigue strength of the material and in the outcome of non-destructive inspection entails the use of reliability based optimization in order to identify an adequate maintenance scheduling.

Most contributions on this area apply the so-called First Order Reliability Method. However, the First Order Reliability Method may be inaccurate in case the performance function is strongly nonlinear or in case it is a high dimensional problem. In this contribution, the evaluation of the reliability is performed by means of advanced simulation methods, in particular, by means of Subset Simulation. Advanced simulation methods have been successfully applied in structural dynamics and stochastic finite elements and to fatigue analysis.

In this work, a numerical strategy for designing an optimal maintenance scheduling for a structure, accounting explicitly for the effects of uncertainty is suggested. This contribution, which can be regarded as an extension of the methods developed in, presents several novel aspects over similar approaches proposed in the literature. Firstly, the initiation and propagation of fatigue crack is modeled efficiently by means of cohesive zone elements. The application of this class of elements allows modeling the crack initiation and propagation within a unified framework. It should be noted that cohesive zone elements have already been used for uncertainty quantification of the crack propagation phenomenon. However its application within the context of maintenance scheduling constitutes a novelty. The second innovative aspect of this contribution refers to the assessment of the reliability sensitivity with respect to the variables that define the maintenance scheduling. The estimation of this sensitivity, which is required in order to determine the optimal maintenance schedule within the proposed framework, can be quite demanding as the model characterizing repair of a cracked structure leads to a discontinuous performance function associated with the failure probability. A new approach for modeling this function is proposed herein. The continuous and discontinuous parts respectively of the function are considered separately to estimate accurately the gradients of the failure events.

Crack Propagation Phenomenon

Mechanical components may deteriorate under cyclic loadings. One or several cracks may initiate and propagate through the structure, leading to an eventual structural failure of the component, or to a loss of serviceability. The fatigue life is characterized by three different stages: fatigue crack initiation, stable crack growth and unstable crack growth. During the crack initiation stage, damage accumulates at the microscopic level. In the case of a metallic material, one or several micro-cracks initiate at stress concentration points or at the defects of the material (inclusions, grain boundaries, etc.). These micro-cracks progressively grow and coalesce until a macroscopic crack appears. The crack initiation is strongly affected by the micro-structural parameters (size of the inclusions or grain orientation at the stress concentration, etc.) Thus the time to crack initiation depends on parameters that cannot be fully controlled at the macroscopic level and can be modeled as an uncertain process.

The propagation stage is first characterized by stable crack growth. The crack length increases progressively during the fatigue life, and the crack partially propagates through the cross-section of a structural component. The crack propagation stage is also an uncertain process since it is influenced by the microscopic structure. Once the cracks reach a critical size, the cross section of the structure is so reduced that it can no longer sustain the applied load. The structure is partially or fully destroyed by brittle failure or ductile collapse.

The most widely used model to predict fatigue crack growth is expressed by the Paris–Erdogan equation or any of its further implementations. They consist of a phenomenological relation between the crack growth rate and the stress intensity factor range. Numerical methods have been developed in order to determine the stress intensity factor of complex structures incorporating one or several cracks, such as the extended finite element method. This method can be used in combination with the Paris–Erdogan equation to model fatigue crack growth. However, specific requirements have to be met to ensure that Paris–Erdogan equation is predictive. The crack must exhibit a certain minimum initial length and the yielding at the crack tip should not be excessive. However, these conditions do not apply to most engineering structures.

Cohesive zone elements are an alternative method to account for crack growth by means of finite element simulation. Such models have been pioneered by Dugdale and Barrenblatt. They consist of zero-thickness elements that are inserted between the bulk elements and account for the resistance to crack opening by means of a dedicated traction-displacement law. This cohesive force dissipates, at least partially, the energy related to crack formation.

Unfortunately, the cohesive zone elements as described above are not suitable for modeling fatigue crack growth. In such cases, the stiffness of the cohesive elements does no longer evolve after few cycles, leading to crack arrest (i.e. the crack length is no longer increasing). Nguyen et al. extended the cohesive law to include fatigue crack growth, which is modeled by the means of a deterioration of the material properties at each cycle. During the unloading–reloading process, the cohesive law shows a hysteresis loop, the slight decay of the stiffness simulates fatigue crack propagation. Such cohesive elements account for both the crack initiation and the crack propagation, respectively. The crack growth phenomenon is modeled in this contribution using cohesive zone elements. The uncertainties inherent in the fatigue crack initiation and propagation are modeled by means of random variables (grouped in a vector θ) for the material parameters of the cohesive zone elements. Thus, the uncertainty in these material parameters propagates to the crack initiation and propagation phenomena.

Modeling of Non-destructive Inspection

The deterioration of mechanical components subject to fatigue leads to a decrease of reliability. In order to ensure sufficient reliability during lifetime, two different strategies may be adopted:

- The reliability completely relies on the design of the structure, involving appropriate sizing of the components, quality assurance of the parts during manufacturing or with the use of conservative safety factors.

- Sufficient reliability is maintained by a program of periodic inspections, which allows to assess the service conditions of the structure. The damaged components can be replaced, repaired or strengthened when necessary, which guarantees an extended service life or a less costly design.

The selection of one of these strategies depends on the service conditions and on costs considerations. The second strategy is suitable for components that can be easily accessed and replaced. Furthermore, the application of this second strategy requires the definition of a particular inspection technique and also its periodicity. Several inspection techniques are available to evaluate the degradation of aging structures. The most common ones are visual inspection, penetrant inspection, eddy current, radiographic inspection, ultrasonic inspection, etc. Each method shows particular benefits and also drawbacks. For instance visual inspection can be easily performed and does not require many tools, however it relies mainly on the skills of the inspector and hence human errors cannot be avoided. Radiographic inspection can efficiently detect cracks with a low risk of error, but this method requires costly equipment investments.

All the non-destructive inspection techniques show variability in their outcome. The results are affected by the conditions of inspections, e.g. the flaw size, the geometry of the structure, the particular inspection technique, the inspectors skills, etc. The uncertainties inherent in the non-destructive inspection techniques can be modeled within the framework of probabilistic approaches. In particular, the probability of detection of a crack can be regarded as dependent on the inspection technique and on the flaw size. Several formulations of the probability of detection function have been proposed. Typically, the probability of detection increases with the crack length and reaches a limit value, which might be less than one, for instance because of human errors during the non-destructive inspection.

In this contribution, the probability of detection is modeled with an exponential distribution. The parameter of the distribution is assumed to depend on the respective quality of inspection:

$$POD(l(t,\theta),q) = 1-\exp(-q \cdot l(t,\theta)),$$

where POD denotes the probability of detection, $l(t,\theta)$ denotes the crack length (that depends on time t and the vector of random variables (θ)) and q is the scalar value modeling the quality of inspection. Using equation $POD(l(t,\theta),q) = 1-\exp(-q \cdot l(t,\theta))$, the probability of detection of short cracks is very low, and increases as the crack length $l(t,\theta)$ increases. The parameter q describes the characteristics of the non-destructive inspection. As a matter of fact, an increase of the value of the parameter q corresponds to increased chances of detecting a crack of a given length. For instance, in case $q = 1$ mm^{-1}, the probability of detecting a crack with a length of 1 mm is approx. 63%. In case $q = 5$ mm^{-1}, the probability of detecting a crack with the identical length is approx. 99%. Hence, in this model, selection of the value of q is regarded as equivalent to choosing a particular inspection technique.

The crack length estimated during the non-destructive inspection is affected by sizing errors, i.e. the measured crack size is different from its true size. Several sources of uncertainties may cause the measurement errors the lack of repeatability of the inspection procedure, an inadequate calibration of the measurement device, the geometry of the flaw, the influence of the temperature and humidity, etc. The error in sizing is modeled using a Gaussian distribution, the measured crack length is the sum of the actual crack length and the sizing error.

Life-time Events and Effects of Maintenance

During the lifetime of a structure, different events may occur, i.e. due to crack growth, inspection and eventual repair may be performed. In order to illustrate these events, assume a structure with a single crack where inspection is performed at the time t_I and the critical crack length (i.e. the crack length at which the structure collapses) is denoted as l_c. Thus, the following sequences of events, illustrated schematically in figure, may occur:

- Fracture occurs before inspection.

- In case the structure has not failed before time t_I, non-destructive inspection is performed.

Two situations may occur depending on the outcome of the inspection:

- The structure is not repaired, either because the structure is not jeopardized by the level of damage or because of detection errors. Fracture may or may not occur before the end of the service life at a time $t > t_I$. For case 2 illustrated in figure, fracture does not occur.

- In case the structure is repaired, imperfect removal is considered (i.e. another crack may initiate and grow at the same location). As previously, the crack may lead to failure before the end of the service life or the structure may survive.

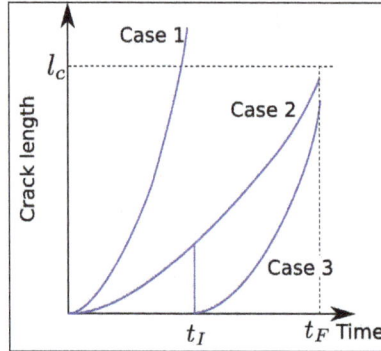

Aspect of the evolution of the crack length, with a maintenance operation.

The specific repair activity to be performed on a structure is problem dependent. For example, the cracked parts of a system can be replaced. In metallic structures, another possible strategy consists of welding the cracks. Alternatively, a patch can be applied to the structure, which consists of a metallic or composite plate glued on the damaged area and which partially carries the load.

Figure below summarizes the events which may happen during the service life of a structure. The repair event and the fracture event are not fully correlated. For instance, the structure may fail before the end of the service life, even though it has been repaired. Indeed, imperfect removal is considered (i.e. a crack may initiate and propagate after repair) and multiple site damage may happen (i.e. several cracks may propagate through the structure and some of them may not be repaired). The structure can as well be safe even though it has not been repaired during its service life.

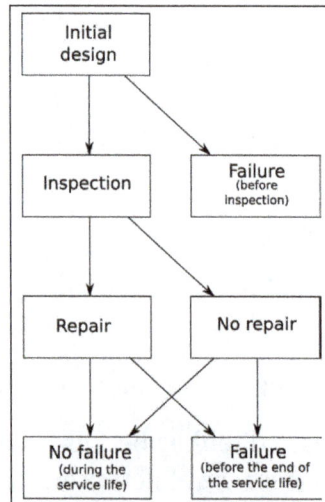

Event tree associated for the service life
of a component, with a single inspection time.

According to the event tree illustrated in figure, two notable events may take place:

- The structure may not be repaired during the maintenance activities, leading to failure before the end of the service life. Such event is caused for instance by the uncertainties inherent in the non-destructive inspection (i.e. the probability of detecting a crack is not equal to one), or by inadequate scheduling of the maintenance activities (i.e. if the maintenance activities are planned too early in the service life, the cracks may be too short to be detected).

- The structure may be repaired although it is not required (i.e. the structure does not fail during its service life without maintenance activities). This situation may be caused by a too conservative maintenance scheme.

The two events described above constitute outcomes that are undesirable. This is because the first event implies failure even though efforts on inspection are being performed while the second event implies an unnecessary repair effort.

As an additional remark, it should be noted that due to the effects of repair, a crack can be removed. Thus, the crack length l does not depend solely on time t and the random variables θ associated with the crack propagation process but also on the time of inspection t_I and the quality of inspection q. Thus, $l = l(t, x, \theta)$, where $x = (q, t_I)^T$.

Formulation of the Performance Functions

In order to characterize the occurrence of the repair and failure events for structural reliability analysis, the so-called performance function is defined with respect to the random variables. The value of this function is less than or equal to zero for those realizations of θ that cause the event of interest (either repair or failure) and larger than zero otherwise.

The performance function is frequently expressed as the difference between the capacity and the demand functions. The performance functions are defined as the difference between a normalized capacity and a normalized demand, as shown in equation $g_X(x, \theta) = 1 - d_X(x, \theta)$, The normalized capacity is equal to one and the normalized demand $d_X(x, \theta)$ is a dimensionless function expressed in terms of the random variables:

$$g_X(x, \theta) = 1 - d_X(x, \theta),$$

where x is the vector of variables defining the maintenance scheme (recall that $x = (q, t_I)^T$), X denotes the life-time events associated with the structure, e.g. the failure or the repair event. d_X and g_X denote the normalized demand and the performance function associated with the event X, respectively, and θ denotes the uncertain parameters.

The performance function associated with fatigue prone components is typically expressed with respect to the actual fatigue life and the target fatigue life (referred to as t_c

and t_F, respectively, in figure below. Following an approach similar to the one developed in, the performance function is expressed with respect to the crack length at the end of the service life l_F and the critical crack length l_c. In case fracture occurs before the end of the service life, the crack is *artificially* propagated beyond its critical length, as depicted in figure below. Clearly, this does not possess any physical meaning. Nonetheless, the cracks are propagated beyond their physical limit as a means for formulating the performance functions associated with repair and failure, respectively. For instance, in case repair is not considered, it suffices to check the crack length at the time t_F in order to determine whether fracture occurs, instead of checking the crack length for any time instant $t \in [0, t_F]$. The artificial crack length increases with a constant rate with time, which is equal to the crack growth rate at the last cycle before fracture occurs.

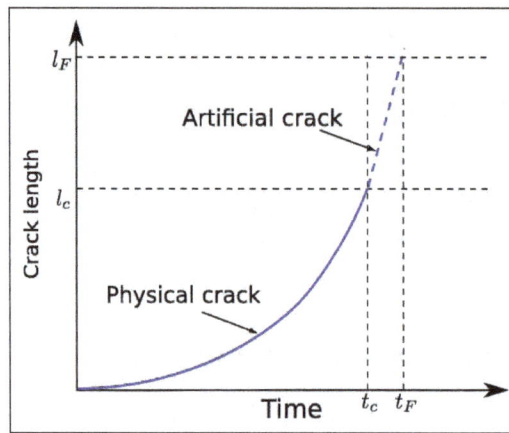

Introduction of an artificial crack.

Performance Function Associated with the Repair Event

In the numerical model, the decision to repair a crack is taken in case the following requirements are fulfilled:

- The crack is detected during inspection at time t_I. The uncertainties inherent in the crack detection procedure are modeled using an extra random variable θ_d with an uniform distribution in the range $[0, 1]$. The non-destructive inspection fails in detecting the crack if $\theta_d \geq \text{POD}(l(t_I, x, \theta))$, otherwise the crack is detected. This formulation leads to the detection of a crack of the length $l(t_I, x, \theta)$ with a probability equal to $\text{POD}(l(t_I, x, \theta))$.

- The structure is repaired only if the measured crack length $l_{meas}(t_I, x, \theta)$ exceeds a given threshold length l_{th}. This is equivalent to performing repair if $l_{meas}(t_I, x, \theta)/l_{th} \geq 1$. The estimation of the crack length is affected by measurement errors, which is modeled with an additive variable: $l_{meas}(t_I, x, \theta) = l(t_I, x, \theta) + \epsilon$, where ϵ denotes the error in sizing of the crack, its value may be greater than zero (the crack length is overestimated) or less than zero (the crack length is underestimated). In this contribution, ϵ is modeled by a Gaussian distribution.

- The structure is repaired in case fracture has not occurred before the inspection time, which is equivalent to having $l_c(x, \theta)/l(t, x, \theta) \geqslant 1, t \in [0,t_I]$, where $l_c(x, \theta)$ denotes the critical crack length. Once the crack length reaches $l_c(x, \theta)$, unstable crack growth occurs, which propagates through the whole structure during a cycle. A crack is artificially propagated beyond its critical length. Thus, the condition of no failure before the inspection time can be checked by means of the inequality $l_c(x, \theta)/l(t_I, x, \theta) \geqslant 1$.

Each crack of the structure is repaired if the three conditions stated are fulfilled. Hence, the associated normalized demand associated with a crack $d_{R,i}$ is defined as:

$$d_{R,i}(x,\theta) = \min\left(\frac{POD(l_i(t_I,x,\theta),q)}{\theta_{d,i}}, \frac{l_{meas,i}(t_I,x,\theta)}{l_{th,i}}, \frac{l_{c,i}(x,\theta)}{l_i(t_I,x,\theta)} \right)$$
$$i = 1...N_C,$$

where the subscript i refers to the N_C cracks present in the structure, $\theta_{d,i}$ is the uncertain parameter associated with crack detection, $l_i(t, x, \theta)$ is the actual crack length, $l_{th,i}$ threshold crack length at which the decision of repair is taken, $l_{c,i}(x, \theta)$ is the critical crack length.

In case repair actions are taken, all the cracks that fulfill the three requirements stated above are removed from the model. The other cracks have identical lengths before and after the inspection.

Repair actions may be necessary in case at least one of the cracks fulfills the requirements stated above and the associated normalized demand is expressed as:

$$d_{R,*}(x,\theta) = \overset{max}{_i}(d_{R,i}(x,\theta)), \quad i = 1...N_C,$$

where $d_{R,*}$ denotes the performance function associated with the repair of one of the cracks. The structure is not repaired in case fracture occurs before the time of inspection, i.e. in case the length of one of the cracks exceeds its critical value before the time of inspection t_I, which is expressed as:

$$d_{F,tI}(x,\theta) = \overset{max}{_i}\left(\frac{l_i(t_I,x,\theta)}{l_{c,i}(x,\theta)} \right), \quad i = 1...N_C,$$

where $d_{F,tI}$ denotes the normalized demand associated with fracture before the time of inspection t_I.

Subsequently, the performance function associated with the repair event d_R is expressed as:

$$d_R(x,\theta) = \overset{min}{_i}\left(d_{R,*}(x,\theta), \frac{1}{d_F,t_I(x,\theta)} \right), \quad i = 1...N.$$

Performance Function Associated with Fracture

Failure occurs during the service life if the length of one of the cracks $l_i(t_I, x, \theta)$ exceeds a critical value $l_{c,i}(x, \theta)$, thus leading to unstable crack growth. The normalized demand associated with failure can be expressed as:

$$d_F(x,\theta) = \overset{max}{i}\left(\frac{l_i(t_I,x,\theta)}{l_{c,i}(x,\theta)}, \frac{l_i(t_F,x,\theta)}{l_{c,i}(x,\theta)}\right), \quad i=1\ldots N_C,$$

where t_I is the inspection time and t_F is the target life time. Note that the normalized demand function introduced in equation,

$$d_F(x,\theta) = \overset{max}{i}\left(\frac{l_i(t_I,x,\theta)}{l_{c,i}(x,\theta)}, \frac{l_i(t_F,x,\theta)}{l_{c,i}(x,\theta)}\right), \quad i=1\ldots N_C,$$

checks the occurrence of failure at two specific times only instead of checking failure at each time $t \in [0, t_F]$. Nonetheless, this strategy is still valid due to the fact that in this contribution, cracks are artificially propagated beyond their critical length and that the crack length is a function increasing with time. Thus, the failure condition can be still captured by equation,

$$d_F(x,\theta) = \overset{max}{i}\left(\frac{l_i(t_I,x,\theta)}{l_{c,i}(x,\theta)}, \frac{l_i(t_F,x,\theta)}{l_{c,i}(x,\theta)}\right), \quad i=1\ldots N_C,$$

regardless failure occurs at some time t different from t_I or t_F.

Design of a maintenance scheduling by means of reliability-based optimization:

As stated previously, the effects of uncertainties cannot be neglected for scheduling of maintenance activities. Uncertainties are considered in the non-destructive inspection, as well as in the crack initiation and growth processes. Hence, the costs associated with repair and fractures are not fixed, but they are influenced by the uncertain parameters. The optimum of a function including uncertainties can be found in the framework of reliability based optimization. Several definitions of reliability-based optimization have been proposed in the literature. The outcomes from reliability analysis can be considered in the performance function, or in the constraints, or in both. Herein, a function whose expression includes a linear combination of outcomes from reliability analysis is minimized. The problem of reliability based optimization is formally stated as:

$$\underset{x=(q,tI)^T}{\min} \quad C_T(x),$$

Subject to $h_i(x) \leqslant 0, \quad i=1\ldots N_c,$

where C_T denotes the total life time costs of the structure, which have to be minimized, $h_i(x)$ denote the constraint functions, which are fulfilled as long as their value is less than (or equal to) zero and N_c is the total number of constraints. The time of inspection t_I and quality of inspection q are introduced as the design variables of the optimization procedure (i.e. the objective of the study is finding the values of these parameters leading to minimized total costs). Only deterministic constraint functions are considered herein (i.e. they do not depend on the outcome of a reliability analysis).

In this study, total costs are expressed as the summation of the costs of inspection, repair and failure:

$$C_T(x) = C_I(x) + C_R(x) + C_F(x),$$

where C_I, C_R and C_F denote the cost functions associated with inspection, repair and failure respectively. Following the approach developed in no additional information about the relative costs is considered and it is assumed that the there is a linear relation between the costs associated with the uncertain events (fracture, repair) and their respective probability of occurrence. Similarly, the costs associated with inspection are assumed to be proportional to the parameter q. The proportionality coefficients weigh the different events (inspection, repair and fracture) according to their contribution to the total costs.

The costs associated with inspection are assumed to be proportional to the quality of inspection:

$$C_I(x) = C_i \cdot q$$

where C_i is a coefficient weighting the contribution of the inspection to the total costs.

The costs associated with repair and failures are expressed as:

$$C_R(x) = C_r \cdot p_R(x)$$

$$C_F(x) = C_f \cdot p_F(x)$$

where p_R and p_F are the probability of repair and the probability of fracture during the service life respectively, C_r and C_f are coefficients weighting the contribution of the repair and of the failure of the structure within the total costs respectively. In the formulation of equation $C_R(x) = C_r \cdot p_R(x)$, the number of repaired cracks does not affect the costs associated with the repair activities.

Within the scope of this manuscript, the outcome of an inspection is used to decide whether or not repair should be carried out. Hence, the information collected at inspection time is used solely for deciding the most appropriate time for inspection and also the best strategy for performing that inspection (which is related to the quality parameter). In other words, the problem is designing an optimal maintenance schedule for a generic mechanical component subject to fatigue damage. However, it is important to note that the outcome of an inspection can be also used for updating the reliability of a particular structure by means of, e.g. Bayesian approaches. That is, for a structure that has been built and where one has some prior knowledge on its state involving fatigue damage, the information gathered by inspection activities may allow updating the knowledge on the state of the component and taking decisions on repair for that particular structure. The latter approach is outside the scope of this contribution.

Solution Strategy

Modeling of Fatigue Cracks using Cohesive Elements

Fatigue cracks are expected to initiate at the rivet holes and propagate through the structure until fracture occurs.

The use of cohesive zone elements allows to treats cracks by means of finite element simulation. They consist of zero-thickness elements that are inserted between the bulk elements and account for the resistance to crack opening using a specific traction-displacement law. The cohesive force dissipates, at least partially, the energy related to crack formation. The use of such elements to account for fracture has been pioneered by Dugdale and Barenblatt. In this context, the crack growth is seen as a gradual phenomenon, with the progressive separation of the lips of an extended crack.

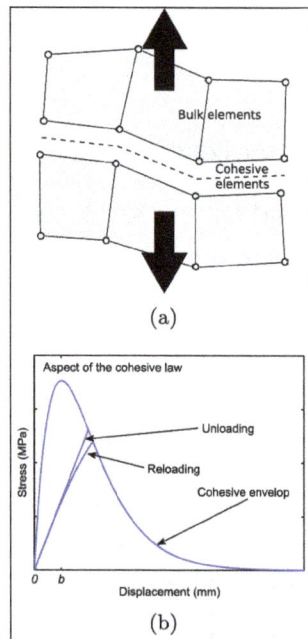

(a) Insertion of cohesive zone elements at the interface of bulk
elements. (b) Aspect of the traction-displacement law for cohesive elements.

Nguyen et al. extended the cohesive law to include fatigue crack growth. If the classical cohesive elements are used to model a cracked body undergoing alternating stress, the parameters of the finite element model do no longer evolve after few cycles, leading to crack arrest. The effects of the history are modeled using deterioration of the stiffness with time. During the unloading–reloading process, the cohesive law shows a hysteresis loop. A slight decay of the stiffness is introduced to simulate fatigue crack propagation. This approach has been successfully used to model fatigue crack growth.

Using the principle of virtual work, the mechanical equilibrium of a solid containing a cohesive surface can be expressed as:

$$\int_V \sigma : \delta\varepsilon dV - \int_{Sint} T_{coh} \cdot \delta\Delta dS = \int_{Sext} T_{ext} \delta u dS,$$

where V, S_{int} and S_{ext} are the bulk volume, the cohesive and external surface respectively, σ, T_{coh} and T_{ext} denote the stress tensor, the cohesive traction vector and the external traction vector respectively, $\delta\varepsilon$ is the symmetric gradient of the test displacement field u. Δ denotes the relative displacement between adjacent cohesive surfaces. The second term of the left-hand side of equation $\int_V \sigma : \delta\varepsilon dV - \int_{Sint} T_{coh} \cdot \delta\Delta dS = \int_{Sext} T_{ext} \delta u dS$, represents the contribution of cohesive elements to the total mechanical energy.

The resistance of a material to crack formation can be expressed considering the energy dissipated during the formation of a new surface within the material. The total amount of energy dissipated during the formation of this surface is expressed as the sum of the energy related to destroying the chemical bonds between the atoms (or molecules) constituting the material and the energy associated with the plastic strain at the vicinity of the interface (e.g. the energy associated with the crack tip plasticity):

$$\Gamma_s = \Gamma_d + \Gamma_p,$$

where Γ_s denotes the total amount of energy associated with the creation of the interface, Γ_p is the energy associated with plastic strain and Γ_d is the energy associated with debonding. Regarding cohesive zone element models, the plastic strain in the bulk elements at the crack tip accounts for Γ_p and the traction-displacement law dedicated to the cohesive elements accounts for Γ_d.

During a finite element simulation, stable crack growth occurs as long as the mechanical energy associated with the boundary condition can be dissipated by the elements. When this energy can no longer be dissipated, unstable crack growth occurs. In this contribution, the critical crack length is defined as the crack length at the last instant before fracture.

The mechanical model proposed by Needleman is used in the case of monotonic loading. The cohesive stress is expressed as:

$$T_n = a \cdot \delta_n \cdot \exp\left(-\frac{\delta n}{b}\right),$$

where δ_n denotes the displacement of the opposite nodes of an element in the normal direction, T_n is the normal stress within a cohesive element, a and b are material parameters. When such an element undergoes separation, the cohesive force first increases, which models the resistance of material to crack propagation. If the displacement exceeds a critical value, the cohesive force decreases, which accounts for the loss of strength of the damaged material (i.e. voids or micro-cracks appear in front of the crack tip).

Equation $T_n = a \cdot \delta_n \cdot \exp\left(-\dfrac{\delta n}{b}\right)$, does not apply though when unloading is considered. Indeed, the behavior of the cohesive elements has to account for the irreversibility of crack growth. The stiffness of the cohesive elements is reduced by damage and unloading occurs linearly at constant stiffness so that stress vanishes when the separation is equal to zero.

In conventional formulations of cohesive zone elements, an unloading–reloading cycle is performed at constant stiffness values. Such formulations are applicable to fracture mechanics only. The cohesive law, as presented up to now, is non-dissipative, since there is no degradation of the material properties over a cycle, leading to crack arrest after few cycles. The material law proposed in equation $T_n = a \cdot \delta_n \cdot \exp\left(-\dfrac{\delta n}{b}\right)$, is extended to cyclic loading in the implementation of the cohesive zone element. The material law consists of a cohesive envelope describing the behavior of an element under monotonic loading and a hysteresis loop accounts for the damage accumulation at each fatigue cycle. When a cohesive element undergoes unloading and then reloading, the stiffness decreases slightly as the stress is increased. The loss of stiffness of damaging material can be assessed with a scalar damage parameter D whose value is within the range [0–1]. Several authors used such a scalar parameter in the context of cohesive zone elements under fatigue loadings. The rate of loss of stiffness is expressed as:

$$\frac{dD}{dt} = \alpha \cdot Tn(t)^\beta \cdot \max(T_n(t) - T_0, 0)^\gamma,$$

where D is the total damage accumulated within an element, α, β, γ and T_0 are material parameters and t is the time. The parameter T_0 is the stress at which damage does no longer accumulate within the material. In case of homogeneous repartition of the stress (at least among the crack path), the fatigue limit is equal to the value of T_0. The coefficient α monitors the rate at which damage accumulates. The coefficients β and γ monitor the sensitivity of damage rate to the stress.

At any instant during cyclic loading, the stress in a cohesive element is equal to:

$$Tn(t) = \frac{a}{b} \cdot (1 - D(t)) \cdot \delta_n,$$

where a and b are the parameters of the cohesive envelop law, given by equation,

$$T_n = a \cdot \delta_n \cdot \exp\left(-\frac{\delta n}{b}\right),$$

δ_n is the relative displacement in the direction normal to the center-line of the element.

In case of monotonic loading, the traction-displacement law is that of the cohesive envelope. Unloading of a structure can be defined as a decrease of the applied stress. However, this definition cannot be systematically generalized to the behavior of one single cohesive element. Local unloading can be caused by global unloading of the structure, by a change in the repartition of stress as a crack propagates or by interactions between cracks. Since cohesive elements show softening, loading (resp. unloading) is defined as an increase (resp. decrease) of the separation (i.e. displacement of opposite nodes of the element).

The case $D = 0$ corresponds to virgin material. When the first loading is applied, the behavior of the element is determined by equation $T_n = a \cdot \delta_n \cdot \exp\left(-\frac{\delta n}{b}\right)$, until unloading occurs. The case $D = 1$ corresponds to completely damaged elements, which do not transfer any stress. Such elements correspond to the physical crack.

The value of the damage parameter is equal to one in the elements at the crack location and its value decreases progressively with the distance from the crack. However, there is a progressive transition between the cracked material and the uncracked material. The elements with a damage parameter greater than 0.99 are assumed to be fully damaged, and a cohesive crack is defined as a succession of adjacent fully damaged elements.

At the beginning of the fatigue life, all the cohesive zone elements have a damage parameter $D = 0$, and the model does not include any crack. After the first load cycles, the value of the damage parameter D increases faster at the elements located nearby the stress concentration zones (e.g. near a sharp angle, a hole, etc.). As long the damage parameter is smaller than one the cohesive elements account for crack propagation.

A crack is introduced in the finite element model when an element is fully damaged ($D = 1$). The stress concentration zones migrate at the newly formed crack tip, causing a faster increase of the damage parameter in the elements near the crack tip. At this stage, the cohesive zone elements account for fatigue crack propagation. Once the crack reaches a critical length, the cohesive elements can no longer compensate the stress concentration at the crack tip and fracture occurs. As stated before, in this contribution, the critical crack length l_c is defined as the length of the crack obtained just before fracture.

The growth of fatigue cracks is considered in the aluminum alloy 2024-T3. The plasticity of the bulk elements is modeled using the Voce law. Table shows the values of the parameters of bulk material and of the cohesive envelop (coefficients a and b of equation $T_n = a \cdot \delta_n \cdot \exp\left(-\dfrac{\delta n}{b}\right)$, determined by fitting the data.

Table: Material properties used in finite element simulations.

Variable	Value	Unit
Young's modulus	70000	MPa
Poisson ratio	0.3	–
Yield stress	330	MPa
Ultimate stress	650	MPa
Coefficient a of equation $T_n = a \cdot \delta_n \cdot \exp\left(-\dfrac{\delta n}{b}\right)$	1500	MPa
Coefficient b of equation $T_n = a \cdot \delta_n \cdot \exp\left(-\dfrac{\delta n}{b}\right)$	0.05	mm

Stochastic crack growth can be modeled using correlated random variables in order to model the coefficients of the equations governing fatigue crack growth.

The coefficients α, β and γ are modeled with fully correlated random variables. A previous study showed that the coefficient α of equation,

$$\frac{dD}{dt} = \alpha \cdot Tn(t)^{\beta} \cdot \max(T_n(t) - T_0, 0)^{\gamma},$$

can be modeled by a random variable θ_α using a lognormal distribution and the coefficients β and γ can be modeled by a Gaussian distribution, as indicated in table.

Table: Value of the parameters monitoring crack growth, defined in equation:

$$\frac{dD}{dt} = \alpha \cdot Tn(t)^{\beta} \cdot \max(T_n(t) - T_0, 0)^{\gamma}.$$

Variable	Type	Mean	Standard deviation
α	Lognormal	1.9×10^{-5}	3.2×10^{-5}
β	Gaussian	0.21	0.0083
γ	Gaussian	0.42	0.017
T_0	Deterministic	100	–

Meta-modeling

The finite element simulation using cohesive zone elements is extremely demanding from a computational viewpoint. Three factors contribute to the computational

time associated with the numerical simulation of fatigue crack growth using cohesive elements:

- The formulation is strongly non-linear. Hence, several inversions of the tangent
  matrix are required to model the behavior of a structure over one fatigue cycle.

- It is necessary to repeat a large number of simulations of the behavior over one
  single cycle in order to describe accurately the behavior. The simulations can be
  accelerated by the means of special algorithms. However, it is necessary to repeat
  many times the finite element simulations of the behavior of the structure over
  an individual cycle in order to model accurately the fatigue crack growth. Most of
  the computational efforts are spent on these successive simulations over a cycle.

- Most of the fatigue life is spent during the crack initiation or during the growth
  of short cracks. In order to accurately model these processes, the finite element
  mesh must be refined at the crack initiation sites.

The use of meta-models (or surrogate models), such as response surface models, Gaussian process or Kriging interpolation allows to approximate the crack length or the fatigue
life with limited computational efforts. The use of meta-models is well adapted to reliability analysis, which requires a large number of computations of the performance function.

In this study, linear regression is used to approximate the outcomes of time consuming finite element simulations. A set of N_{reg} independent basis functions
$T_{reg} = \{T_{reg,1},...,T_{reg,Nreg}\}$ is selected. The meta-model is expressed as:

$$\widehat{F}(t,\theta,l_0,B) = \sum_{j=1}^{N_{reg}} B_j \cdot T_{reg,j}(t,\theta,l_0) + e_{reg},$$

where $\widehat{F}$ denotes the response surface, $T_{reg,j}, j = 1 ... N_{reg}$ denotes the basis functions
used in the regression, e_{reg} is the regression error. The regression variables consist of
the time t the uncertain parameters θ and the initial crack lengths l_0. During a simulation of the fatigue life, no crack is initially present in the model and the terms of l_0 are
all equal to zero. The consideration of initial cracks allow to model fatigue crack growth
after repair activities, in case cracks are removed from the model. $B = \{B_1,...,B_{Nreg}\}$ is the
vector of the regression parameters, which has to be determined in order to minimize
the regression error.

The least square estimate $\widehat{B}$ of the regression parameters can be expressed as:

$$\widehat{B} = (X^T X)^{-1} Xy_{full} \text{ model,}$$

where $y_{full\ model}$ denotes the set of outcomes of the finite element simulation corresponding to the training points, X is a matrix containing the value of the basis functions for
the different values of the training points, i.e.

$$X_{ij} = T_{reg,j}\left(t^{(i)}, \theta^{(i)}, l_0^{(i)}\right), i = 1 \dots N_{SP,j} = 1,$$

where N_{SP} denotes $\dots N_{reg}$ the number of support points and $\left(t^{(i)}, \theta^{(i)}, l_0^{(i)}\right)$ denotes the support points. Response surfaces are calibrated to approximate the actual length of the cracks l_i at any instant of the service life, and the critical length of the cracks $l_{c,i}$. The response surfaces are directly used in the formulation of the performance functions instead of the outcome of the finite element simulations.

Response surfaces approximating the crack lengths in the time range $[t_p, t_F]$ are required. At the instant t_p, the structure may include cracks at some of the initiation sites (these cracks have not been repaired during the maintenance activities). At the other initiation sites, there may be no crack at the instant t_p since repair activities have been performed. In order to approximate accurately the crack lengths in the time range $[t_p, t_F]$, training points with initial cracks are considered in order to calibrate the response surface. However the initial crack lengths l_0 are not included in the model of uncertainties, since cohesive zone elements account for fatigue crack initiation.

Efficient method allows to use meta-models in order to perform reliability analysis without systematic bias by means of Subset Simulation. However, in the context with reliability-based optimization, the performance function is expressed with respect to the random variables and the design variables, respectively. Hence, it is necessary to calibrate as well as meta-model accounting for the random variables and the design variables, respectively.

Assessment of Reliability

Reliability analysis aims at determining the probability that a component reaches a given state condition. In this study, the state conditions of interest are respectively repair and failure of the structure.

The uncertain parameters are modeled with random variables and the probability can be expressed through the following multidimensional integral:

$$p(x) = \int_{g(x,\theta)<0} f(\theta)\, d\theta,$$

where θ denotes the uncertain parameters, f is the joint probability density function and g represents the performance function. Reliability analysis can be performed e.g. by means of Monte Carlo simulation, that consists of generating samples of the random variables and counting the number of outcomes within the failure region:

$$\hat{p}(x) = \frac{1}{N}\sum_{i=1}^{N} I_f(x, \theta^{(i)}),$$

where $\hat{p}$ is the approximation of the failure probability, N is the number of samples generated, $\theta^{(i)}$ denotes the samples and I is the indicator function, which is equal to one for the samples in the failure region and zero elsewhere. However, Monte Carlo simulation generally requires to generate a very large number of samples, which is computationally prohibitive when small failure probabilities (e.g. 10^{-6}) have to be estimated.

The advanced procedure of Subset Simulation allows to estimates small failure probabilities with a limited number of evaluations of the performance function. It is based on a decomposition in intermediary failure events. A set of intermediary failure regions is defined so that $F_1 \supset F_2 \supset \cdots \supset F_m$, where F_m is the failure region whose probability of occurrence has to be determined. The probability associated with the intermediary failure region can be estimated with limited computational efforts. The final failure probability can be determined by conditional probabilities:

$$p \simeq P(F_1)\prod_{i=1}^{m-1} P(F_{i+1} \mid F_i),$$

where $P(\cdot)$ denotes the probability associated with an event.

Reliability Sensitivity Estimation

Besides determining the probability of repair and failure, the sensitivities (gradients) of each of these probabilities with respect to time of inspection and quality of inspection are required for determining an optimal maintenance schedule according to the optimization strategy considered. This procedure allows to estimates the gradients of the failure probabilities at reduced computational costs.

The partial derivative of the probability with respect to x_i is defined as:

$$\frac{\partial p}{\partial x_i} = \lim_{\delta x \to 0} \frac{P(g(x+\delta x \cdot \epsilon_i, \theta) \leqslant 0) - P(g(x,\theta) \leqslant 0)}{\delta x}$$

where ϵ_i is a vector with the same size as the set of design variables x, the ith element of ϵ_i is equal to one, the other terms are all equal to zero. Recall the design variables consist of the time and quality of inspection (i.e. $x = (q, t_i)^T$). In order to evaluate the partial derivative of equation $\dfrac{\partial p}{\partial x_i} = \lim_{\delta x \to 0} \dfrac{P(g(x+\delta x \cdot \epsilon_i, \theta) \leqslant 0) - P(g(x,\theta) \leqslant 0)}{\delta x}$ efficiently, two approximations are introduced. First, a local linear approximation of the performance function is performed in the vicinity of the design variables of interest:

$$g(x + \delta x \cdot \epsilon_i, \theta) \in g(x,\theta) + \beta_{o,i} \cdot \delta x,$$

where $\beta_{o,i}$ is a scalar parameter. The procedure for computation of the parameter $\beta_{o,i}$ can be summarized as follows: (i) a subset of samples within the vicinity of the limit state is selected; (ii) the performance function is computed for these samples considering perturbed design variables (i.e. the value of $g(x + \delta x \times \epsilon_i, \theta)$ is determined); (iii) the coefficient $\beta_{o,i}$ is computed from the results of the previous steps, for instance using linear regression.

The second approximation introduced to estimate the partial derivative of equation
$$\frac{\partial p}{\partial x_i} = \lim_{\delta x \to 0} \frac{P(g(x+\delta x \cdot \epsilon_i, \theta) \leqslant 0) - P(g(x,\theta) \leqslant 0)}{\delta x} \text{ is:}$$

$$P\big(g(x,\theta) - \Xi \leqslant 0\big) \approx e^{\alpha_1 + \alpha_2 \cdot \Xi},$$

where α_1 and α_2 are two scalars determined using linear regression, Ξ denotes a perturbation term, which is set to have a local approximation of the probability of interest (for instance, $\Xi \in [-0.1, 0.1]$). At first, the probability in the left hand side of equation $(P\big(g(x,\theta) - \Xi \leqslant 0\big) \approx e^{\alpha_1 + \alpha_2 \cdot \Xi},)$ is computed for various values of Ξ. Then the coefficients α_1 and α_2 are determined using linear regression. This approximation has been used successfully in several publications within the area. If the reliability analysis has been performed beforehand (for instance to estimate the objective function), the coefficient α_1 and α_2 are estimated without extra performance function evaluations. The results obtained from the reliability analysis performed previously are reused and only the count of the samples leading to a performance function value below the threshold level Ξ is computed.

Considering the two approximations described above, it can be shown that the sought partial derivative can be estimated by means of the following expression:

$$\frac{\partial p}{\partial x_i} = -\beta_{0,i} \, \alpha_2 p(x).$$

Sensitivity of the Failure Probability

One of the main assumptions behind the approach described above for estimating the sensitivity of the probability is that the associated performance function is continuous. However, the performance function associated with the failure event may not fulfill this condition. In order to overcome this issue, a strategy is proposed in the following.

It should be noted that the performance function associated with probability of fracture shows discontinuities with respect to the random variables monitoring fatigue crack growth and with respect to the parameters associated with crack detection. Indeed, in the numerical model, a slight variation of one of these parameters may lead to detection and repair of a crack that was initially not repaired (and reciprocally), leading to a discontinuity in the performance function. In order to clarify this issue, consider the following qualitative example. Assume a plate with an edge crack, which undergoes an inspection with perfect sizing of the crack and the probability of detecting the crack may be defined by equation $POD(l(t,\theta),q) = 1 - \exp(-q \cdot l(t,\theta))$.

Uncertainties are considered in the crack growth rate and in the outcome of non-destructive inspection. Figure illustrates the shape of the performance function associated with failure.

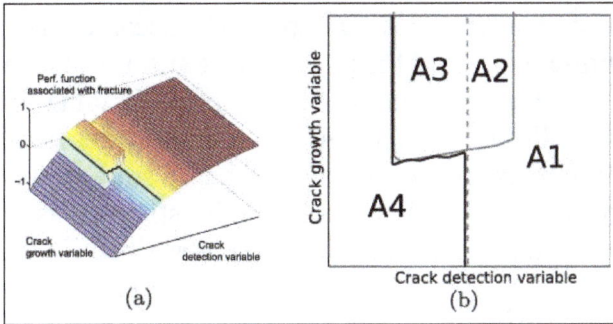

Performance function related to fracture.

(a) General aspect of the performance function. The thick black line denotes the limit state. (b) Division of the space of the random variable in several sets. The thick solid black line denotes the limit state of the performance function related to fracture, the thick solid gray line denotes the limit state of the performance function related to repair, the thick dash gray line denotes the limit state of the performance function associated with failure, in case no maintenance activities are considered.

Figure shows the division of the random variables space in 4 zones. A1 denotes the zone for which the random variables values always lead to a safe structure, structural failure does not occur during the entire service life and the structure is not repaired. The zone A2 denotes the region of values leading to repair of the structure, although the structure is safe without repair. The zone A3 denotes the region of values leading to failure when no maintenance is performed and leading on the other hand to a safe structure if maintenance activities are performed. The zone A4 denotes the region of values leading to failure, either because the crack fails in being detected during the non-destructive inspection, or because the crack reaches its critical length before the time of inspection t_r.

The zones A2 and A3 correspond to the regions of values leading to repair of the structure, and the performance function associated with fracture is discontinuous at the border of these regions. This discontinuity imposes a major challenge when analyzing the sensitivity of the failure event with respect to the different parameters relevant to the model. On the contrary, the discontinuity between the zones A1 and A2 does not affect the analysis, since these zones are both in the safe domain.

In the simple example proposed, the discontinuity is due to the fact that whenever a crack is repaired, its length changes suddenly from a given value to zero (when perfect repair is considered). Imperfect repair is considered, another crack initiates and grows. Its final length is not correlated with the crack length without repair. This sudden change clearly introduces a discontinuity in the associated performance function. In order to cope with the discontinuity discussed previously two artificial performance functions are introduced, which are associated with the subsets of the space of the random variables.

The first function is related to the continuous part of the performance function (i.e. the safe domain consist of area A1 and A2 in figure. The second performance function is related to the discontinuous part of the performance (i.e. the failure domain consist of area A3 in. It can be expressed as the probability of performing necessary repair (i.e. the structure would fail if it is not repaired and it is safe after perfect repair). Both of these performance functions are continuous and hence suitable for sensitivity estimation. The probability of failure (and its gradients) is estimated as the difference of the probabilities defined by the performance functions described above:

$$p_F(x) = p_0 - p_{NR}(x),$$

where p_F denotes the probability of failure (fracture before the target life), p_{NR} defines the probability of necessary repair and p_0 denotes the probability of failure without repair activities, which is not expressed in terms of x (since the structure is not repaired). Hence, the probability p_0 is determined before starting the optimization and its value does no need to be updated at each iteration.

Using equation $p_F(x) = p_0 - p_{NR}(x)$, in case the structure is repaired, the simulation of the life time events have to be performed twice i.e. with and without the repair activities.

The normalized demand associated with the probability of failure without repair d_0 can be expressed as:

$$d_0(\theta) = \max_i \left(\frac{l_i(t_F, \theta)}{l_{c,i}(\theta)} \right) \quad i = 1 \ldots N_C.$$

In this study, the necessary repair operation is defined as fulfilling the following requirement:

- The crack is detected and repaired right after the inspection.

- Fracture occurs before the end of the service life without repair.

The normalized demand associated with the probability of *necessary repair* P_{NR} can be expressed as:

$$d_{NR}(x, \theta) = min(d_0(\theta), d_R(x, \theta)).$$

Optimization Strategy

The objective is to determine the maintenance schedule by minimizing the total costs associated with the maintenance and eventual failure of the structure, the sensitivity associated with the reliability analysis can be determined efficiently. Hence gradient-based optimization algorithms are well suited for solving the reliability based optimization problem. In particular, a first order scheme based on feasible directions is applied in this contribution. This scheme is implemented due to its simplicity and robustness but certainly other optimization schemes based on gradients that are more efficient could be applied as well.

The method of feasible directions involves two main steps. In the first one, for a given a feasible design x_k (i.e. a design fulfilling the constraints of the optimization problem) a search direction d_k is determined such that it is possible to find a sufficiently small step $\xi > 0$ fulfilling the condition $C_T(x_k + \xi d_k) < C_T(x_k)$. The search direction d_k can be determined by solving a linear programming problem involving the gradient of the objective function and the active constraints.

The second step of the method of feasible directions consists in exploring the one dimensional space defined by the search direction d_k, i.e. a line search is performed. The objective is determining an optimal step ξ^{opt} that solves the following one-dimensional optimization problem:

$$\min_{\xi} \quad C\frac{L}{T}(\xi) = C_T(x_k + \xi d_k)$$
$$\text{Subject to } \xi > 0, \; h_i(x_k + \xi d_k) \leqslant 0,$$
$$i = 1 \ldots N_c,$$

where $C\frac{L}{T}(\cdot)$ is the total costs function along the search direction. For solving this one dimensional optimization problem, the step $\bar{\xi}$ to the nearest active constraint is determined using any appropriate search scheme such as bisection. Once $\bar{\xi}$ has been found, the optimal step ξ^{opt} is calculated by means of the following criterion. In case the derivative of $C\frac{L}{T}(\cdot)$ is negative at $\bar{}$, then $\xi^{opt}=\bar{\xi}$ and the new feasible design is $x_{k+1} = x_k + \xi^{opt} d_k$. In case the derivative of $C\frac{L}{T}(\cdot)$ is positive at $\bar{\xi}$, then the optimal step is located in the interval $[0, \bar{\xi}]$. Thus, the value of the optimal step can be determined using again a bisection scheme.

For the actual implementation of the line search step described above, it should be noted that it might be necessary to $C\frac{L}{T}(\cdot)$ several times. As its evaluation is numerically demanding (because it implies calculating probabilities), it is proposed to approximate this function by a polynomial:

$$C\frac{L}{T}(\xi) \approx \bar{C}\frac{L}{T}(\xi) = C_0 + C_1 \xi + C_2 \xi^2,$$

where $C_j, j = 0, 1, 2$ are real coefficients. These coefficients are determined using the values of the cost function and of its sensitivity along the search direction (directional derivative) evaluated at three points $(\xi_1, \xi_1, \xi_3) \in [0, \bar{\xi}]$. It is clear that only the function values at three points would be required for determining the sought coefficients. However, it should be kept in mind that there is an inherent variability associated with the evaluation of the total costs function as it depends on probabilities that are evaluated by means of simulation. Thus, the extra data (directional derivative) improve the robustness of the method by coping, at least partially, with the variability inherent to simulation methods.

Numerical Example

The objective of this example is designing a maintenance schedule for a metallic component subject to cyclic loading. The structure studied consists of a plate with two rivet holes with a diameter of 4 mm each. The plate has a height of 400 mm, a width of 64 mm and a thickness of 2.3 mm. The loading is applied in the longitudinal direction, with a maximum stress of 200 MPa and a minimum stress of 40 MPa. The symmetry of the structure among its center-line in the transverse direction (represented by a dashed line on figure is considered and the finite element model consists of half of the plate. The mesh is refined at the rivet holes in order to describe accurately the repartition of the stress at the rivet holes. The mesh refinement also improves the accuracy of the modeling of fatigue crack initiation and of the propagation of short cracks.

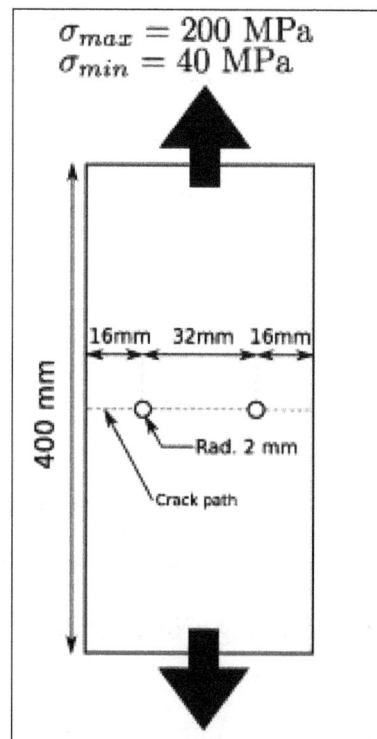

Geometry of the structure.

Cohesive zone elements are inserted at the crack path, as indicated in figure. The uncertainties inherent in the fatigue crack initiation and propagation are influenced by parameters showing spacial variation within the structure (such as the micro structural properties). In case the coefficient of equation $\frac{dD}{dt} = \alpha \cdot Tn(t)^\beta \cdot \max(T_n(t) - T_0, 0)^\gamma$, monitoring the fatigue crack initiation and growth is modeled using a single random variable, the time to crack initiation is the same for the four sites where cracks initiate (at the holes of the structure shown on figure. Thus, all the cracks have the same length at any instant of the service life. This is obviously incorrect, since one could expect to have a single crack initiating from one of the sites and then propagating through the structure. Thus the parameters α, β and γ of equation $\frac{dD}{dt} = \alpha \cdot Tn(t)^\beta \cdot \max(T_n(t) - T_0, 0)^\gamma$, are modeled with spacial variation within the structure. Four independent random variables are used, where each of them is devoted to one of the crack initiation sites. At each extremity of the central ligament, the coefficient α is equal to the realization of the random variable devoted to this crack initiation site. This coefficient shows a linear variation within the central ligament. In each of the ligaments at the extremities of the structure, the coefficient α is constant (i.e. there is no spacial variation within each of the ligaments). The coefficient α is equal to the realizations of the random variable devoted to this location.

The error in sizing of the crack is modeled with Gaussian distribution with zero mean and a standard deviation equal to 2.4×10^{-3} mm. The sizing error may be different for each of the cracks present in the structure. Hence, four independent random variables are used in the model. Similarly, the independent random variables are used in the formulation of the probability of detecting a crack, described in equation $d_{R,i}(x, \theta).....i = 1...NC$. The uncertainties inherent in the detection, sizing, initiation and propagation of the cracks are modeled using 12 random variables in total.

The structure has a target fatigue life of 250,000 cycles. The structure is considered safe if fracture does not occur, i.e. none of the cracks has reached the critical length leading to unstable crack growth. One inspection activity is considered during the total life. It is assumed that the coefficient q can assume any (real) value. The threshold crack length $l_{th,i}$ is equal to 1 mm. It is assumed that the cracks with a length below this value do not jeopardize the structure and are not repaired after the inspection, even though these cracks may be successfully detected. The same threshold length is used for all the cracks of the model.

The coefficient related to the costs of inspection, repair and failure are equal to $C_i = 5 \times 10^{-3}$, $C_r = 2.5$ and $C_f = 100$, expressed in arbitrary monetary unit. The objective of the reliability based optimization is minimizing the total costs. The side constraints for the design variables are $140,000 \leqslant t_I \leqslant 250,000$ and $1 \leqslant q \leqslant 30$.

For launching the optimization procedure, the maintenance schedule is selected such that the inspection is performed after 148,000 cycles and the coefficient q is equal to 28.6 mm⁻¹.

In case a crack initiates at the side of a rivet hole, it is likely to have another crack emanating from the opposite side of the hole. Proppe and Schuëller modeled the initiation of cracks emanating from the same rivet hole with correlated random variables. However, such approach is not required herein. The parameter a monitors the initiation and the growth of the cracks, and is modeled with independent random variables (at each site of crack initiation). However, the lengths of the cracks at the different sites are actually correlated. This correlation is caused by the repartition of the stress in the structure in the presence of cracks. As an example, when a crack appears at a side of a hole, the stress at the opposite side is increased, which speeds up the initiation and growth of a crack at this location.

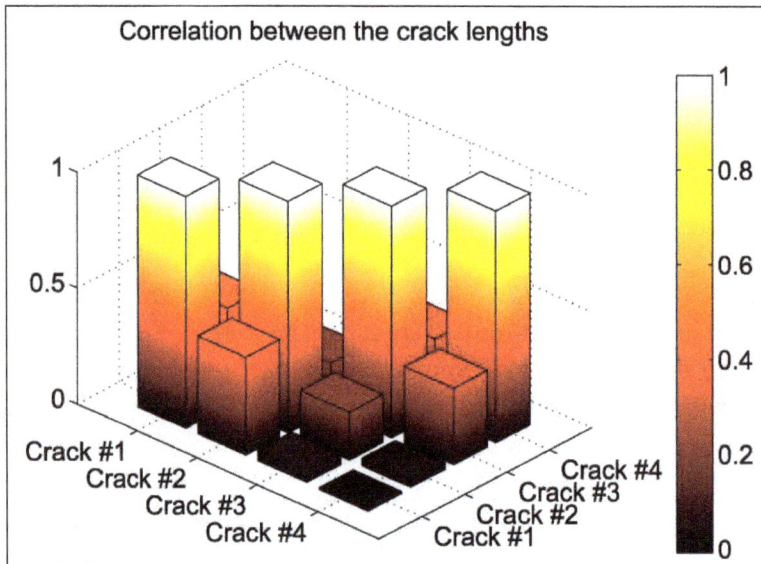

Correlation matrix between the cracks lengths after 200,000 cycles, .

Figure show the costs associated with fracture, repair and inspection respectively as a function of the time of inspection and quality of inspection. The costs associated with inspection increase linearly with the quality of inspection respectively. The costs associated with repair are strongly affected by the time of inspection. Indeed, the latter the inspection is performed, the longer the cracks are in the structure, which require repair, causing the increase of the associated costs. The costs of repair are slightly affected by the quality of inspection, which increase the chances of detecting cracks. The costs associated with fracture are strongly affected by the time of inspection. In case the inspection is performed too early, the likelihood of detecting a crack is very low, and the decision to repair the structure is not taken. In case the inspection activities are performed too late, the probability of failure before the inspection is rather high, which leads to an increase of the associated costs.

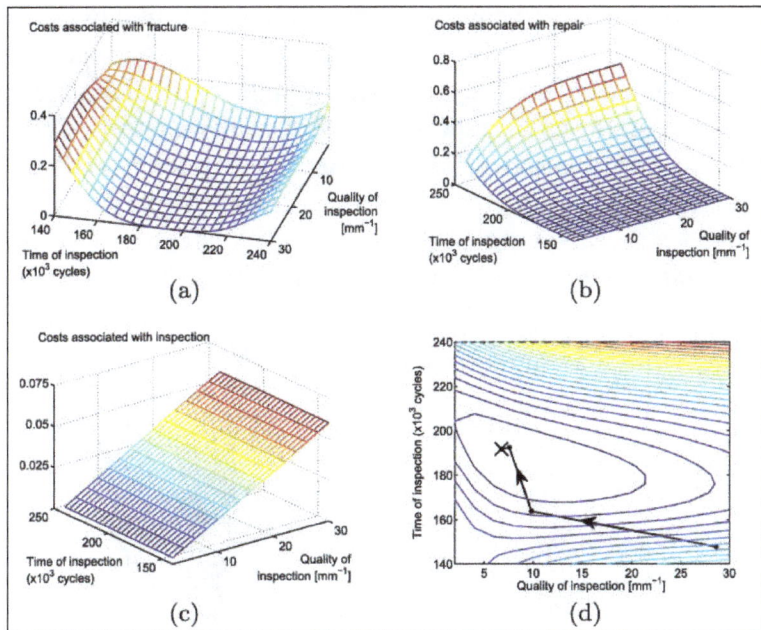

Costs associated to the model.

(a) Costs associated with fracture. (b) Costs associated with repair. (c) Costs associated with inspection. (d) Evolution of the design variables during the reliability based optimization procedure. The contour lines show the total costs (in arbitrary monetary units), the solid lines show the successive search directions, the dots represent the intermediary designs, the cross shows the coordinates of the optimum.

The total costs are shown in figure above by means of contour lines. The function of the total costs shows one minimum, and is relatively flat at the vicinity of its minimum. The same figure illustrates the trajectory of optimization algorithm in the space of the design variables. The details on each point along this trajectory are summarized in table. The procedure converges efficiently towards the optimum. At the first iteration, the total costs are greatly reduced. The procedure arrives to the vicinity of the optimum and the costs are further reduced at subsequent iterations. The minimum costs could be found after three iterations. In total, three line searches were necessary, which represents 10 successive runs of Subset Simulation. The total computer time required to perform reliability sensitivity (computation of the gradients) is negligible when compared to the computational time associated with reliability analysis.

Table: Value of the inspection parameters during the optimization procedure.

Iteration	q (mm^{-1})	$t_I \times 10^3$ Cycles	Total costs
Initial design	28.6	148	0.51
Intermediary design 1	9.8	163	0.21
Intermediary design 2	7.6	192	0.11
Final design	6.7	191	0.09

In addition to the information provided in figure and table, the first two columns of figure provide details on the costs associated with inspection, repair and failure for the initial design and optimal design, respectively. It is seen that the optimal maintenance schedule is a compromise between the costs associated with these three events. The initial maintenance strategy is not appropriate and the costs related with failure are the dominant ones. As the optimization progresses and the optimal maintenance schedule is found, the costs associated with repair increase, but this allows a subsequent decay of the costs associated with failure, leading to a decrease of the total costs associated to the structure.

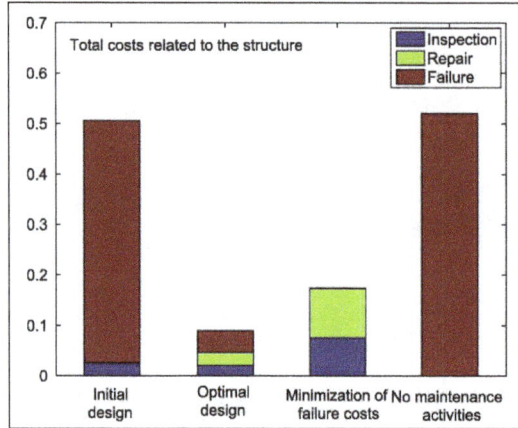

Total costs associated with the structure.

Considering the optimal maintenance scheduling, the inspection is performed late during the service life. Indeed, the optimal value of the time of inspection is equal to 191,000 cycles, which corresponds to approx. 76% of the service life of the structure. The structure is not damaged at the beginning of its service life, and the amount of damage progressively increases.

In order to gain insight about the tradeoff that arises between inspection, repair and maintenance costs when looking for an optimal maintenance schedule, two additional cases were analyzed. The first case involves minimizing the costs of failure alone and determining the associated optimal maintenance schedule. Then, the total costs associated for that optimal maintenance schedule are calculated and plotted in the third column of figure. The second additional case studied corresponds to calculating the total costs when no maintenance activities are considered. As no maintenance activities are considered, the total costs for this second case are equal to the failure costs. This last result is plotted in the fourth column of figure. It is most interesting to note that minimizing the failure costs alone leads to total costs that are considerably larger than the case where all costs are considered for optimization (second column of the figure). In addition, it can be noted that suppressing maintenance activities (fourth column of the figure) causes a dramatic increase of the total costs. Details on the costs associated with the second, third and fourth columns of figure are summarized in table. These results highlight the importance of considering all costs when searching for an optimal maintenance schedule, as the optimal solution is evidently a tradeoff between different factors.

Table: Costs associated with different maintenance strategies.

Case	q (mm^{-1})	$t_I \times 10^3$ Cycles	Inspection costs	Repair costs	Failure costs	Total costs
Minimization of total costs	6.7	191	0.03	0.02	0.04	0.09
Minimization of failure costs	15.8	198	.08	0.1	1.2×10^{-3}	0.18
No maintenance activities	-	-	0	0	0.52	0.52

A method for determining optimal maintenance scheduling of metallic structures considering uncertainties has been proposed herein. Cohesive zone elements provide a framework to investigate fatigue crack growth. Contrary to approaches based on linear fracture mechanics, cohesive elements do not require to introduce explicitly initial cracks. The degradation associated with cyclic load is modeled by means of an internal damage parameter which increases during the fatigue life. Cracks appear once the elements are fully damaged. This approach accounts for fatigue crack initiation and propagation using the same phenomenological model. The variability inherent in fatigue of a structure has been assessed using a stochastic model for the parameters monitoring the evolution of the damage. The uncertainties related to fatigue crack initiation and to crack propagation are accounted for using a single model for uncertainties. Moreover, the model describing the crack detection includes its inherent variability. It is assumed that the outcome of non-destructive inspection can be fully represented by its probability of detection.

The performance function associated with fracture is discontinuous, which is not suitable for the estimation of reliability sensitivity. The gradients have been estimated by introducing two auxiliary performance functions, one of them is accounting for the continuous part of the gradient, the second one is accounting for the effects of the discontinuities.

The methods presented here allowed to find the optimal schedule for the maintenance activities. The time and quality parameter of the inspection leading to the minimum costs associated with the structure were determined. The evaluation of the cost function over a grid showed that the costs associated with such structure are mainly affected by the time of inspection and in less degree by the quality of inspection.

The computational efforts are greatly reduced by introducing a meta-model (e.g. a response surface), using an advanced simulation method for the reliability analysis and an efficient algorithm for computing the gradients of the failure probabilities.

Concerning the numerical example and the results obtained, it is most interesting to observe that the determination of an optimal maintenance schedule with respect to total costs implies finding a tradeoff between the costs of inspection, repair and eventual failure. Thus, it is not sufficient to consider one of these three events by itself, as it may lead to a suboptimal scheduling of maintenance activities.

References

- Public-notesse3, public-notes: lysator.liu.se, Retrieved 21 July, 2019

- What-is-preventive-maintenance-optimisation-pmo, reliability-improvement: assetivity.com.au, Retrieved 15 January, 2019

- Planned-maintenance-optimization, maintenance-tools, learning: onupkeep.com, Retrieved 3 May, 2019

- Optimize-preventive-maintenance, works: emaint.com, Retrieved 13 February, 2019

5

Maintenance Management

The maintenance activities are planned, organized and monitored by using administrative and technical framework called maintenance management. It is used as a quality process for the management of human errors. This chapter has been carefully written to provide an easy understanding of maintenance management.

Maintenance management is all about maintaining the resources of the company so that production proceeds effectively and that no money is wasted on inefficiency.

Maintenance of facilities and equipment in good working condition is essential to achieve specified level of quality and reliability and efficient working. Plant maintenance is an important service function of an efficient production system. It helps in maintaining and increasing the operational efficiency of plant facilities and thus contributes to revenue by reducing the operating costs and increasing the effectiveness of production.

Maintenance Objectives

Maintenance in any activity is designed to keep the resources in good working condition or restore them to operating status.

The objectives of plant maintenance are:

- To increase functional reliability of production facilities.

- To enable product or service quality to be achieved through correctly adjusted, serviced and operated equipment.

- To maximize the useful life of the equipment.

- To minimize the total production or operating costs directly attributed to equipment service and repair.

- To minimize the frequency of interruptions to production by reducing breakdowns.

- To maximize the production capacity from the given equipment resources.

- To enhance the safety of manpower.

Maintenance Costs

Breakdown of equipment makes the workers and the machines idle resulting in loss of production, delay in schedules and expensive emergency repairs. These downtime costs usually exceed the preventive maintenance costs of inspection, service and scheduled repairs up to the point M shown in figure.

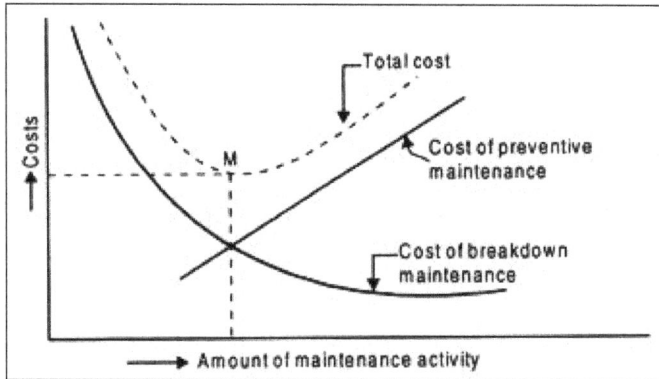

Maintenance Costs.

Beyond this optimal point an increasingly higher level of preventive maintenance is not economically justified and it is economical to adopt breakdown maintenance policy. The optimal level of maintenance activity M, is easily identified on a theoretical basis, to do this the details of the costs associated with breakdown and preventive maintenance must be known.

Costs associated with maintenance are:

- Down time (Idle time cost) cost due to equipment breakdown.

- Cost of spares or other material used for repairs.

- Cost of maintenance labour and overheads of maintenance departments.

- Losses due to inefficient operations of machines.

- Capital requirements required for replacement of machines.

Maintenance Policies

- Breakdown (repair) maintenance.

- Preventive maintenance.

Breakdown maintenance is emergency based policy in which the plant or equipment is operated until it fails and then it is brought back into running condition by repair. The maintenance staffs locate any mechanical, electrical and any other fault to correct it immediately. Preventive maintenance policy prevents the probable breakdown and it ensures smooth and uninterrupted production by anticipating the breakdowns (failures) and taking corrective actions:

- The preventive maintenance policy has four forms:

  ○ Time based: Which means doing maintenance at regular intervals? It is time dependent rather than usage dependent.

  ○ Work based: Maintenance after a set of operating hours of volume of work produced.

  ○ Opportunity based: Where repair and replacement takes place when the equipment or system is available.

  ○ Condition based: Which often relies on planned inspection to reveal when maintenance is required? Preventive maintenance is used to delays or prevents the breakdown of equipment and also to reduce the seventy of any breakdowns that occur.

- Two aspects of preventive maintenance are:

  ○ Inspection: Inspection of critical parts will indicate the need for replacement or repair well in advance of probable breakdown. Regular inspection conducted by either by equipment or operator or by maintenance department is the most important direct means of increasing equipment reliability.

  ○ Servicing: Routine cleaning, lubrication and adjustment may significantly reduce wear and hence prevent breakdowns. Frequently such duties are carried out by equipment operator or may be carried out by maintenance department.

Preventive versus Breakdown Maintenance

Preventive maintenance is the routine inspection and service activities designed to detect potential failure conditions and make minor adjustments or repairs that will help prevent major operating problems.

Breakdown maintenance is the emergency repair and it involves higher cost of facilities and equipment that have been used until they fail to operate.

Effective preventive maintenance programmes for equipment requires properly trained personnel, regular inspection and service and has to maintain regular records.

Preventive maintenance is planned in such a way that it will not disturb the normal operations hence no down time cost of equipment. Breakdown maintenance stops the normal activities and the machines and the operators are rendered idle till the equipment is brought back to normal condition of working.

Maintenance Performance

The following criteria can be used for measuring the effectiveness of maintenance function:

$$1.\text{Productivity of maintenance} = \frac{\text{Output (product)}}{\text{Maintenance cost (Product)}}$$

$$2.\text{Down time index} = \frac{\text{Downtime hours}}{\text{production hours}} \times 100$$

$$3.\text{Maintenance cost index} = \frac{\text{Maintenance cost}}{\text{Capital cost}} \times 100$$

Importance, Objectives and Functions

Necessity of Maintenance Management

Maintenance activities are related with repair, replacement and service of components or some identifiable group of components in a manufacturing plant so that it may continue to operate at a specified 'availability' for a specified period.

Thus maintenance management is associated with the direction and organisation of various resources so as to control the availability and performance of the industrial unit to some specified level. Thus maintenance management may be treated as a restorative function of production management which is entrusted with the task of keeping equipment/machines and plant services ever available in proper operating condition.

The minimization of machine breakdowns and down time has been the main objective of maintenance but the strategies adopted by maintenance management to achieve this aim have undergone great changes in the past. Maintenance has been considered just to repair the faulty equipment and put them back in order in minimum possible time.

In view of the utilization of mostly general purpose/conventional machines with low production output, the demands on maintenance function were not very high. But with fast developments in the design, development and mechanisms of control such as electronic, NC and CNC in machine tools the manufacturing scenario has changed a lot.

The stringent control of dimensional tolerances and surface finish of the product have increased the tendency to adopt standardization and interchange-ability of parts/components of machines.

In the current production setups even a minor down time leads to serious production problems both technological as well as economical. All this is due to tough competition in the industrial market. Under the present circumstances effective and objectively designed efforts to update maintenance management has become a necessity.

Importance of Maintenance Management

Maintenance management is responsible for the smooth and efficient working of the industrial plant and helps in improving the productivity. It also helps to keep the machines/equipment in their optimum operating conditions. Thus plant maintenance is an important and inevitable service function of an efficient production system. It also helps in maintaining and improving the operational efficiency of the plant facilities and hence contributes towards revenue by decreasing the operating cost and improving the quality and quantity of the product being manufactured.

As a service function it is related with the incurrence of certain costs. The important component of such costs are — employment of maintenance staff, other minor administrative expenses, investment in maintenance equipment and inventory of repair components/parts and maintenance materials.

Absence of plant maintenance may lead to frequent machine breakdown and failure of certain productive centres/services which in turn would result in stoppages of production activities, idle man and machine time, dislocation of the subsequent operations, poor quality of production, failure to meet delivery dates of product supply, industrial accidents endangering the life of workers/operators and allied costs etc.

However, the importance of plant maintenance varies with the type of plant and its production but it plays a prominent role in production management because plant breakdown creates problems such as:

- Loss of production.

- Rescheduling of production.

- Materials wastage (due to sudden stoppage of process damages in process materials).

- Need for overtimes.

- Need for work subcontracting.

- For maximum manpower utilization workers may need alternative work due to temporary work shortages.

Hence, the absence of planned maintenance service proves costlier. So it should be provided in the light of cost benefit analysis.

The main objectives of maintenance management are as follows:

- Minimizing the loss of productive time because of equipment failure to maximize the availability of plant, equipment and machinery for productive utilization through planned maintenance.

- To extend the useful life of the plant, machinery and other facilities by minimizing their wear and tear.

- Minimizing the loss due to production stoppages.

- To ensure operational readiness of all equipment's needed for emergency purposes at all times such as fire-fighting equipment.

- Efficient use of maintenance equipment's and personnel.

- To ensure safety of personnel through regular inspection and maintenance of facilities such as boilers, compressors and material handling equipment etc.

- To maximize efficiency and economy in production through optimum utilization of available facilities.

- To improve the quality of products and to improve the productivity of the plant.

- To minimize the total maintenance cost which may consist of cost of repairs, cost of preventive maintenance and inventory costs associated with spare parts/materials required for maintenance.

- To improve reliability, availability and maintainability.

Functions of Maintenance Management

The important functions of maintenance can be summarized as follows:

- To develop maintenance policies, procedures and standards for the plant maintenance system.

- To schedule the maintenance work after due consultation with the concerned production departments.

- To carry out repairs and rectify or overhaul planned equipment/facilities for achieving the required level of availability and optimum operational efficiency.

- To ensure scheduled inspection, lubrication oil checking, and adjustment of plant machinery and equipment.

- To document and maintain record of each maintenance activity (i.e., repairs, replacement, overhauls, modifications and lubrication etc.).

- To maintain and carry out repairs of buildings, utilities, material handling equipment's and other service facilities such as electrical installations, sewers, central stores and roadways etc.

- To carry out and facilitate periodic inspections of equipment and facilities to know their conditions related to their failure and stoppage of production.

- To prepare inventory list of spare parts and materials required for maintenance.

- To ensure cost effective maintenance.

- To forecast the maintenance expenditure and prepare a budget and to ensure that maintenance expenditure is as per planned budget.

- To recruit and train personnel to prepare the maintenance workforce for effective and efficient plant maintenance.

- To implement safety standards as required for the use of specific equipment or certain categories of equipment such as boilers, overhead cranes and chemical plants etc.

- To develop management information systems, to provide information to top management regarding the maintenance activities.

- To monitor the equipment condition at regular intervals.

- To ensure proper inventory control of spare parts and other materials required.

In terms of plants operations the functions of maintenance are:

- The plant must be available as and when required.

- The plant must not breakdown during actual operation state.

- The plant must operate in an efficient manner at required level of plant operation.

- The down time must not interfere with production runs.

- The down time due to breakdown should be a minimum.

To accomplish these conditions there must be complete cooperation and mutual understanding between maintenance and production departments. There must be an effective maintenance policy for planning, controlling and directing all maintenance activities.

The plant maintenance department must be well organized, adequately staffed sufficiently experienced and adequate in number to carry out corrective and timely maintenance with the efforts in minimizing breakdowns.

Maintenance Management as a Quality Process

Maintenance practices and technologies have evolved to meet the needs of the changing industrial environment. The function has evolved from a community of reactive fixers, to dedicated craftsmen, to proactive professionals. The next generation of personnel could well be based on practitioners of Quality Management Systems (QMS).

Maintenance progress can be best demonstrated by its ability provide assurances for reliability. The model for quality Assurance, which meets the same requirements for

Equipment Reliability, is demonstrated in the ISO 9001:2000 Standard, through process-centeredness.

Process-centered management is a system that manages organizational activities as a process. The process is managed through QMS, which is clear on PDCA as a process method. PDCA is Plan, Do, Check Act. Most modern maintenance management activities are not linked to QMS, which have particular management characteristics. Using these characteristics transforms modern maintenance practice into what may be the next generation of maintenance management. The structure of maintenance management as a process model is significant, as an International Standard to adopt as a relevant Standard or best practice for maintenance management's thrust towards reliability.

However, in order to resolve elements of the process the applicable definitions must addressed in order to provide the basis for generating reliability.

Maintenance can be defined as the degradation management of engineered materials (equipment and systems) to retain their performance within their designed operating parameters.

Just as stress can accelerate deterioration of metals in a corrosive environment, operational stress moves equipment and systems toward failure. Limiting stresses within the operating environment maintains reliability.

The elements of the maintenance which are relevant to PDCA are:

- Protecting components from stress.

- Monitoring their condition.

- Undertaking component(s) replacement prior to the failure threshold level caused by stress excesses.

The components of this system are preventive maintenance (PM), condition monitoring (CM), and planned overhaul (PO). It is against this background that maintenance activities are identified.

Preventive Maintenance

Preventive maintenance is whatever action is undertaken to provide protection against or reduce the rate of degradation by:

- The application of coatings, adjustments, or cleaning to retain components within design operating conditions.

- The physical or chemical control of applied media to the operating environment, which has the potential to increase the stress in components beyond their design operating parameters.

Condition Monitoring

- Condition monitoring identifies the adequacy of preventive maintenance for the stresses of the environment.

- Condition monitoring identifies the rate of degradation towards the failure threshold level.

- Condition monitoring is measurement of degradation or the rate of degradation.

Condition monitoring plays a dual role in the maintenance process for regulating preventive maintenance applications and, as stress increases, alerting to impending failure. Another school thought is that condition monitoring is sometimes viewed not as a part of the maintenance process, but merely as a provider of process measurement for the "check' activity of the process.

Planned Overhaul

Planned overhaul takes place prior to failure, when the rate of degradation is excessive as noted by condition monitoring, and the component(s) is approaching the threshold failure level. It involves the disassembly, component replacement, and re-assembly of equipment.

The maintenance attributes of PMs CMs and POs, is repeatable for equipment as it goes through its life cycle.

The characteristics of Quality Management Systems (QMS) are as follows as it applies to maintenance:

- The 'Plan' to achieve reliability is based within the context of PMs, CMs and OH at the component level, while defining through documentation, the human, economic and technical resources to achieve the reliability.

- The 'Do' is the actual work instructions to be performed and the options at the point of execution for informing of the requirements and record keeping of its reliability status.

- The 'Check' is the means to identify whether the reliability requirements are being met within the context of the human, economic and technical resources.

The measurement of efficiency and effectiveness of the plan determines whether waste through errors or unwanted activity has occurred. It can track the use of resources to determine when they are beyond those proposed by the plan for reliability. Condition monitoring techniques are also applied as relevant measurements toward assured reliability.

These measurement parameters provide the trigger at the analysis stage to determine the component status for likely improvements toward reliability. In addition, the utilization of Pareto Analysis based on formatting root causes by harsh environment, improper operation, recommended end of life or lack of maintenance, will resolve improvement issues required for efficient and effective maintenance management.

The 'Act' is any action taken as a result of checking to provide the feedback mechanisms to the Plan for adjusting any of the PMs, CMs, or POs to ultimately reach the reliability levels required with the context the stated resources.

The application of the PDCA process above contains the characteristics of all Quality Management Systems, which provides performance assurances or reliability:

- Feedback,

- Measurement,

- Record keeping,

- Procedures/Work Instructions,

- Resource Identification.

Maintenance Management systems with these characteristics provides the methodology for continual improvement towards intended reliability.

The adoption of the QMS model for maintenance has a number of advantages. The model:

- Identifies maintenance resources, as it facilitates its contribution (cash flows, current ratios, etc.) to enhanced organizational management.

- Represents the opportunity for maintenance to adopt similar process centeredness, as the rest of the organization, in the event QMS forms part of the management of the organization.

- Determines the lowest cost of reliability proactively. Establishes a methodology for the best practices for Reliability.

- Establishes the selection basis for computerized maintenance management systems.

- Develops zero-based or activity budgets for maintenance; the lowest cost for reliability is measurable and variances can be addressed.

- Establishes a process through QMS for treating maintenance management

- Channels differing methodologies used by maintenance professionals into a world-class standard.

- Establishes a structure and accountability for measuring outsourced maintenance contracting.

- Establishes a method for determining maintenance human resource requirements where reliability assurances can be achieved.

- Defines the specific technical requirements for reliability.

- Establishes the basis for just-in-time (JIT) spares inventory management.

- Determines a methodical application of condition monitoring data and its role in the process.

- Identifies a parallel costing system for maintenance. Provides the basis for pro-actively establishing the contribution of maintenance to product cost.

- Develops root cause analysis for Pareto Analysis for continuous improvement

The ISO 9001: 2000 Scope is customer focused with respect to the effective application of the QMS system. The context is relevant as the operations/production department can be seen as the customer of the maintenance department. Promoting reliability through an improvement process, driven by the characteristics of QMS maintenance management, meets the requirements for the International Standard.

Finally TPM, CBM, RCM etc may very well have the same attributes as QMS, the aim should be for the maintenance community to dovetail all into one Standard for the practice of maintenance management. They all have the same objective of reliability - align them to the Internationally recognized QMS Standard and the egos of 'ownership' and which is better, can go away.

Managing Human Error in Maintenance

Numerous research studies have shown that over 50% of all equipment fails prematurely after maintenance work has been performed on it. In the most embarrassing cases, the maintenance work performed was intended to prevent the very failures that occurred. In their ground-breaking work that led to the establishment of the technique that we now know as Reliability Centred Maintenance, Nowlan and Heap found, when analysing the failures of hundreds of mechanical, structural and electrical aircraft components.

The interesting finding, is that more than two-thirds of all components demonstrated early-life failure. It has been estimated that maintenance errors ranked second to only controlled flight into terrain accidents in causing onboard aircraft fatalities between 1982 and 1991 (despite the application of RCM techniques in the airline industry during this period). A study of coal-fired power stations indicated that 56% of forced outages occur less than a week after a planned or maintenance shutdown.

Other studies have been conducted which confirm these findings, but, until recently, there has been little research performed that has investigated the reasons for this.

Several plausible theories have been proposed – possible explanations that have heard include:

- "Human Error" – The repair/replace task was not successfully completed due to a lack of knowledge or skill on the part of the person performing the repair.

- "System Error" – The equipment was returned to service after a high-risk maintenance tasks without the repair having been properly inspected/tested.

- "Design Error" – The capability of the component being replaced is too close to the performance expected of it, and therefore lower capability (quality) parts fail during periods of high performance demand. The remaining higher capability (quality) parts are capable of withstanding all performance demands placed on it. This could be envisaged in the following graph:

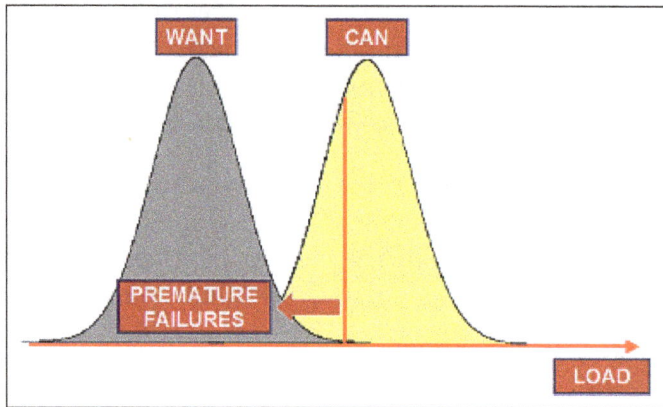

- "Parts Error" – the incorrect part or an inferior quality part has been supplied.

More recently, James Reason has compiled a table summarising the results of three surveys – two performed by the Institute of Nuclear Power Operations (INPO) in the USA, and one by the Central Research Institute for the Electrical Power Industry (CRIEPI) in Japan. In all three of these studies, more than half of all identified performance problems were associated with maintenance, calibration and testing activities. In comparison, on average only 16% of problems occurred while these power stations were operating under normal conditions.

Reason also quoted the results of a Boeing Study which indicated that the top seven causes of inflight engine shutdowns (IFSDs) in Boeing aircraft were as follows:

- Incomplete installation (33%).

- Damaged on installation (14.5%).

- Improper installation (11%).

- Equipment not installed or missing (11%).

- Foreign Object Damage (6.5%).

- Improper fault isolation, inspection test (6%).

- Equipment not activated or deactivated (4%).

We can see from this that only one of these causes was unrelated to maintenance activities, and that maintenance activities contributed to at least 80% of all IFSDs.

If poor quality maintenance causes so many incidents in highly regulated and hazardous industries such as Nuclear Power Generation and Civil Aviation, what proportion of failures may be being caused by Maintenance within *your* organisation? What are the outcomes of maintenance-induced failures? Clearly, depending on the industry in which you operate, there are potentially significant safety and environmental risks. There is a long list of catastrophic failures in which, the inadequate performance of a maintenance task played a significant role. Some of these include:

- Flixborough,

- Three Mile Island,

- Piper Alpha,

- American Airlines Flight 191,

- Bhopal,

- Japan Airlines Flight 123,

- Clapham Junction etc.

But besides the obvious safety risks, perhaps the bigger consequences are economic. General Electric has estimated that each in-flight engine shutdown costs airlines in the region of US$500,000. What could maintenance-induced failures be costing your organisation?

Clearly, we need to do something to reduce the number of equipment failures that are being caused, not prevented, by maintenance. This paper suggests that the most appropriate approach is:

- Admit that human error is inevitable (even in Maintenance) and design our systems and processes around this inevitability.

- Use appropriate tools to ensure that we are not unnecessarily over-maintaining plant and equipment (and therefore increasing the risk associated with the fact that this work may not be performed correctly).

- Work to improve the quality with which maintenance activities are performed – including error-proofing where possible.

Human Error is Inevitable

Think of the traditional engineering approach to dealing with maintenance error, and most engineers tend to think along two lines - either discipline/counsel/train the individual(s) involved, and write a new procedure/work instruction to make sure that it doesn't happen again. Unfortunately, recent research and experience by Behavioural Psychologists indicates that neither of these approaches are likely to be successful in eliminating maintenance error.

Work by Reason and Hobbs explains why maintenance activities can be particularly error-provoking. In particular, it argues the futility of trying to change the human condition, when a more effective way of managing maintenance error is to treat errors as a normal, expected, and foreseeable aspect of maintenance work, and therefore, manage maintenance error by changing the conditions under which that work is carried out.

Reason and Hobbs identified a number of physiological and psychological factors which contribute to the inevitability of human error. These include:

- Differences between the capabilities of our long-term memory and our conscious workspace. In particular, what we call "attention" is closely linked with the activities of the conscious workspace, and the conscious workspace has extremely limited capabilities including:

    ○ Attention is an extremely limited commodity – if it is drawn to one thing, then it is, by necessity, withdrawn from other competing concerns.

    ○ These capacity limits give attention its selective properties – we can only attend to a very small proportion of the total available sensory data we receive.

    ○ Unrelated matters can capture attention – such as preoccupation with other sensory or emotional demands.

    ○ Attentional focus (concentration) is hard to maintain for any more than a few seconds.

    ○ The ability to concentrate depends strongly on the intrinsic capability of the current object of attention.

    ○ The more skilled or habitual our actions, the less attention they demand.

    ○ Correct performance requires the right balance of attention, neither too much or too little.

- The Vigilance Decrement – It is more common for inspectors to miss obvious faults the longer that they have been performing the inspection. This is particularly the case when the number of "hits" is few and far between.

- The impact of fatigue – This could be due to:

  - Time of day effects – Our daily rhythms ensure that we are more likely to commit errors in the small hours of the morning.

  - Stresses - Physical, social, drugs, pace of work, personal factors.

- The level of arousal – Too much or too little arousal impairs work performance.

- Biases in thinking and decision making. There is no such thing as "common sense". In particular we are subject to:

  - Confirmation Bias – Where we seek information that confirms our initial (and often incorrect) diagnosis of a problem.

  - Emotional Decision Making – If a situation keeps frustrating us, then we tend to move into "aggressive" mode, but this often clouds our better judgement.

As a result of these contributing factors, the types of errors that occur most often in Maintenance include:

- Recognition failures – These include:

  - Misidentification of objects, signals and messages.

  - Non-detection of problem states.

- Memory failures – This includes:

  - Input failure – Insufficient attention is paid to the to-be-remembered item. This in turn can include:

    - Losing our place in a series of actions.

    - The "time-gap" experience.

  - Storage failure – Remembered material decays or suffers interference. Most common in maintenance is the problem of forgetting the intention to do something

  - Output failure – Things we know cannot be recalled at the required time – the "what's his name?" experience.

  - Omissions following interruptions – We rejoin a sequence of actions having omitted certain required steps.

  - Premature exits – We terminate a job before all the actions are complete.

- Skill-based slips. Generally associated with "automatic" routines, these can include:

  ○ Branching errors – Such as intending to drive to the golf course on a weekend, but missing the turnoff, and continuing on towards the office as you would every other day of the week.

  ○ Overshoot errors – Intending to stop at the shops on the way home, but forgetting and continuing home without stopping.

- Rule-based Mistakes. Most maintenance work is highly proceduralised, and consist of many "rules". These can be formally written, or exist only in peoples' heads. Typical rule-based errors include:

  ○ Misapplying a good rule – Using a rule in a situation where it is not appropriate.

  ○ Applying a bad rule – The rule may get the job done in certain situations, but can have unwanted consequences. This is most common when people pick up others' "bad habits".

- Knowledge-based errors: Generally the situation when someone is performing an unusual task for the first time. These need not necessarily be committed by inexperienced personnel.

- Violations – Deliberate acts which violate procedures. These can be:

  ○ Routine violations – Committed in order to avoid unnecessary effort, get the job done quickly, to demonstrate skill, or avoid what is seen as an unnecessarily laborious procedure.

  ○ Thrill-seeking violations – Often committed in order to avoid boredom, or win peer praise.

  ○ Situational violations – Those committed because it is not possible to get the job done if procedures are strictly adhered to.

Think of your own situation – have you never committed an error? For most of us, the consequences of our past errors are relatively minor – but that is largely due to luck, and the situation that we were in at the time. The traditional approach to dealing with human error – counselling and writing a procedure – cannot possibly effectively deal with all of the types of errors listed above. We need a more holistic approach for managing maintenance error, and assuring Maintenance Quality.

Avoid Unnecessary "Preventive" Maintenance

Given the statistics mentioned earlier from Nowlan and Heap's work, and others, it is clear that over-maintaining equipment not only is a waste of time and money, but it also increases the risk of safety and environmental incidents, as well as potentially causing expensive and unnecessary failures.

Techniques based on the application of Reliability Centred Maintenance principles are an extremely effective way of weeding out this unnecessary maintenance, and stream-lining and optimising equipment PM programs.

Our analysis of PM programs in place at our clients has indicated that in almost all organ-isations there is a huge amount of unnecessary routine maintenance being performed. In some situations, fewer than 10% of the existing PM tasks were optimal, and it is not unusual for us to identify that as much as half of the routine maintenance activities were, at best, a complete waste of time. In many cases, the performance of some of these "pre-ventive" maintenance activities were potentially causing equipment failures, rather than preventing them – particularly where these activities involved intrusive, fixed interval in-spections and overhauls. At one major offshore oil and gas platform in Western Australia, a comprehensive review of the Preventive Maintenance program led to a 25% reduction in the amount of routine PM being performed. It also led to a 25% reduction in the amount of Corrective maintenance being performed. Clearly, in this case, a fair proportion of the PM that had previously been performed was actually causing, rather than preventing, failures.

The starting point in eliminating unnecessary routine maintenance lies in ensuring that the need for all these routine maintenance tasks is defensibly justified. This is the objective of Assetivity's Rapid Equipment Strategy Development process. This process is based on RCM principles and has ten steps as outlined below:

- Determine Scope of Analysis.

- Verify Equipment Capability.

- Identify Failure Modes.

- Analyse Failure Modes, Effects and Consequences.

- Select Recommended Maintenance Tasks.

- Identify Additional Improvement Tasks.

- Consolidate Schedules and Integrate with Operational Strategies.

- Gain Approval and Implement Recommended Actions.

- Track Success.

- Beyond RCM and PMO.

However that, if you have not already done so, a critical review of your PM program is an essential first step to managing the impact of human error in maintenance.

Maintenance Quality Management – Key Principles

Following Reason and Hobbs, the following are the principles that a Maintenance Quality Management system must embrace:

- Human error is both universal and inevitable: Human error is not a moral issue – making them is as much a part of human life as eating and breathing

- Errors are not intrinsically bad. Success and failure spring from the same roots. We are error-guided creatures. Errors mark the boundaries of the path to successful action

- You cannot change the human condition, but you can change the conditions in which humans work. There are two parts to an error – a mental state and a situation. We have limited control over people's mental states, but we can control the situations in which they have to work.

- The best people can make the worst mistakes: No one is immune to error – if only a few people were responsible for most of the errors, then the solution would be simple, but some of the worst mistakes are made by the most experienced people.

- People cannot easily avoid those actions they did not intend to commit. Blame and punishment is not appropriate when peoples' intentions were good, but their actions did not go as planned. This does not mean, however, that people should not be accountable for their actions, and be given the opportunity to learn from their mistakes.

- Errors are consequences, rather than causes. Errors are the product of a chain of actions and conditions which involve people, teams, tasks, workplace and organisational factors. Discovering a human error is the beginning of the search for causes, not the end.

- Many errors fall into recurrent patterns: More than half of maintenance errors are recognised as having happened before – often many times. Targeting these recurrent errors is the most effective way of addressing human error issues.

- Safety-significant errors can occur at all levels in the system. Indeed, the higher in an organisation that an error is made, the more significant the consequences.

- Error Management is all about managing the manageable. Situations are manageable – human nature, in its broadest sense, is not.

- Maintenance Quality Management is about making good people excellent. Maintenance Quality Management is not about making a few error-prone people better – rather it is a way of making the larger proportion of well-trained and motivated people excellent.

- There is no one best way: Different Maintenance Quality Management methods will apply in different situations, and in different organisations.

- Effective Maintenance Quality Management aims at continuous reform rather than Local Fixes. The temptation is to resolve errors one at a time, as they arise, but as errors tend to be systemic in nature, a more appropriate method is to deal with human error systematically, and continuously.

There are a number of Maintenance Quality Management tools that can be applied. The exact combination of these that is most appropriate for any organisation varies, but they could include:

Person Measures

- Provide training in error-provoking factors: Training maintenance personnel in order to give them an understanding and awareness of the factors and situations that may lead them to be more error-prone is a starting point in successfully addressing human error. They should understand such factors as the limitations of human performance, the limitations of short term memory, the impact of fatigue, the impact of interruptions, the impact of pressure and stress, the types of errors that they can make, and the situations in which these errors are most likely to arise. Once maintainers are aware of their own limitations, then they can start to detect the warning signals that indicate a higher risk of an error being made, and can take steps to avoid this from happening.

- Implement measures to reduce the number of deliberate violations: Traditional approaches to the avoidance of violations tend to focus on scaring people into compliance. This may have its place, but an additional, effective approach is to create a social environment within the workplace where deliberate violations bring disapproval from within peoples' peer groups. There are a number of approaches that are being tried, both within and external to the workplace, which appear to be successfully creating this social environment, but overnight success stories are rare.

- Encourage mental rehearsal of tasks before they are performed: There is significant evidence to suggest that achieving the right degree of mental readiness for a task before it begins has a significant positive impact on the quality and reliability with which this task is performed. This is based on studies of surgeons and Olympic athletes.

- Control Distractions: Anticipating the distractions that are likely to occur, and developing a strategy for dealing with them before they occur is most likely to enhance the quality of task performance.

- Avoid Place-losing Errors: Through such techniques as inserting place-markers at appropriate points in the procedure.

Team Measures

- Provide teamwork training: Significant accidents have occurred as a result of poorly functioning teams. Most notable of these was an aircraft accident involving KLM and PanAm 747s at Tenerife, which resulted in the loss of more than 500 lives. Effective teamwork training will focus on:

  ○ Communication skills.

  ○ Crew development and leadership skills.

  ○ Workload management.

  ○ Technical proficiency.

Workplace and Task Measures

- Ensure that personnel only perform tasks when they are properly trained, skilled and qualified: It goes without saying that quality work practices can only be put in place when maintenance personnel have the requisite technical skills and capabilities required to perform the work that is allocated to them.

- Fatigue Management: Ensure that a well-designed shift roster is in place which minimises the impact of fatigue. Ensure also, that there are adequate controls in place for managing overtime work.

- Assign tasks appropriately: There is evidence to suggest that there is a link between the frequency with which a task is performed, and the likelihood that the task will be performed correctly. Both infrequently performed, and very frequently performed tasks tend to be those at greatest risk of human error. Infrequently performed tasks are generally more at risk because of the lack of experience of the person performing the task, while on very frequently performed tasks fall victim to skill-based slips and lapses, as the person performing the work operates on "auto-pilot". Intelligent allocation of work to individuals takes this into account, and can assist in minimising human error.

- Ensure that equipment, and tasks, are properly designed: In order to minimise the likelihood of error in performing maintenance tasks, the equipment should be designed for maintainability. This should include consideration of such factors as:

  ○ Easy access to components.

  ○ Components that are functionally related should be grouped together.

  ○ Components should be clearly labelled.

  ○ There should be minimal requirement for special tools.

- ◦ It should not be necessary to perform high-precision work in the field.

- ◦ Equipment should be designed to permit easy fault diagnosis.

- Enforce good housekeeping standards: Housekeeping practices are a good indicator of attitudes and culture relating to quality. The correct standards are those that avoid dangerous slovenliness, without resorting to anally-retentive cleanliness.

- Ensure spare parts and tools are managed well: Maintenance cannot perform high quality work if the parts and tools that they need are not available when required. This leads to potentially dangerous short-cuts and workarounds being put in place. An important aspect of Maintenance Quality Management is ensuring that Tool Management and Spare Parts Management processes and practices support the achievement of high quality work.

- Write, and use, effective maintenance work instructions: Omission of necessary steps is the most common form of maintenance error. Some estimates suggest that omissions account for more than half of all human factors problems in maintenance. The development, and use, of effective maintenance work instructions is an important tool in managing these types of errors.

Organisational Measures

- Put in place effective processes for analysing, and learning from, past failures: It is vitally important that any significant failures should be investigated using an effective Root Cause Analysis process. This Root Cause Analysis process, to be effective should fully investigate all of the contributing causes to the failure, whether these be physical causes, human causes, or organisational causes. The most effective solutions to preventing these failures from happening again, will be those that deal effectively with the organisational causes of failures.

  However, in order to effectively analyse those failures that are occurring as a result of human failures, it is also necessary to engender a "Reporting Culture" within the organisation – where all failures, no matter how seemingly insignificant, are reported. This, in turn, particularly when we are dealing with human errors, requires the development of a high level of trust between management and those at lower levels in the organisation. People must not feel that reporting human failures is likely to lead to adverse personal consequences. Those who have researched so-called "High Reliability Organisations" (HROs) have noted that high levels of failure reporting is a significant feature of those organisations.

- Put in place proactive processes for assessing the risk of future maintenance errors: Avoiding the recurrence of past failures is an admirable, but insufficient, goal for those seeking to achieve high quality maintenance outcomes. One possible proactive method that could be employed to proactively manage

Maintenance Quality is to perform a risk assessment of maintenance activities, in order to assess whether the likelihood of human error is high. Possible areas that could be assessed in this risk assessment would include:

- The knowledge, skills and experience of maintenance personnel at all levels.

- Employee morale.

- The availability of tools, equipment and parts to perform maintenance tasks.

- Workforce fatigue, stress and time pressures.

- Shift rosters.

- The adequacy of maintenance procedures and work instructions.

One example of a risk assessment process that is used in the aviation industry is Managing Engineering Safety Health (MESH) which was developed initially by British Airways in the early 1990s, and has been further developed and adapted by Singapore Airlines.

In addition, more specific review and assessment of error detection and containment defences can be performed. This could ask questions such as:

- Are there adequate processes in place for independent inspection of high-risk tasks?

- Are functional tests and checks ever omitted or abbreviated, for any reason?

- Have tasks ever been signed off as completed, when this was subsequently found not to be the case?

- After maintenance, is equipment adequately tested before being returned to service?

Ultimately, even putting both proactive and reactive measures in place will not guarantee the absence of human error, but together, these strengthen the organisation's intrinsic resistance to human error.

References

- Maintenance-management-objectives-costs-and-policies, maintenance-management: yourarticlelibrary.com, Retrieved 17 March, 2019

- Maintenance-management-importance-objectives-and-functions, industries: yourarticlelibrary.com, Retrieved 23 June, 2019

- Maintenance-quality: plant-maintenance.com, Retrieved 21 March , 2019

- Managing-human-error-in-maintenance, maintenance-management: assetivity.com.au, Retrieved 13 July, 2019

6

Maintenance Technologies

Various technologies and systems are used for management and optimization of resources such as intelligent maintenance system, predictive management technologies, computerized maintenance management system, etc. This chapter has been carefully written to provide an easy understanding of these maintenance technologies.

Intelligent Maintenance System

"Intelligent maintenance system" is an umbrella term for a number of approaches that share a common goal: to improve the efficiency of maintenance activities through the use of digital technologies.

An intelligent maintenance system (IMS) uses sensors, hardware processors, cloud applications and advanced analytics to improve the performance of maintenance for a machine, production line or manufacturing facility.

How Manufacturing Facilities benefit from Utilizing an IMS

The challenge in answering the questions above lies in the fact that they all depend on a large number of interlinked parameters; parameters that change over time both in the short and long-term. Fluctuations in environmental factors, the quality of raw materials, asset health status, workforce, market demand, and many others affect the required rate and type of maintenance.

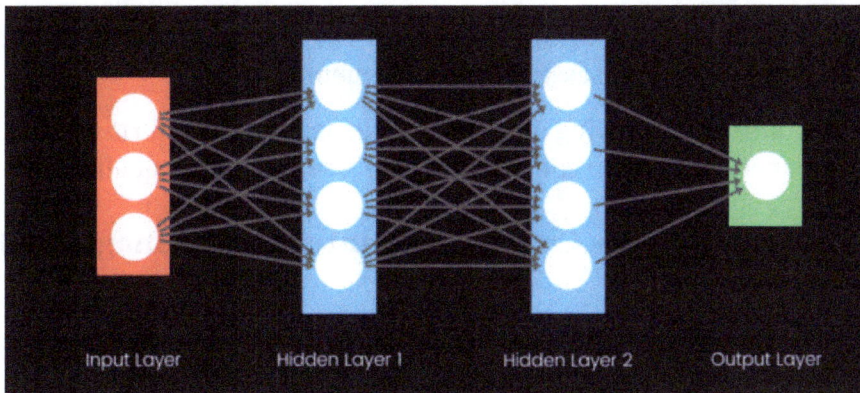

Schematic of an artificial neural network. ANNs are used to discover causal correlations between root causes and failures.

One of Industry 4.0's most powerful use cases is predictive maintenance which leverages data captured from sensors, PLCs, data historians, ERPs, MESs etc. to form failure predictions.

Machine learning and other AI algorithms such as artificial neural networks are used to process this data constantly. The algorithms search for correlations that can help determine the root cause of recurring problems that lead to unplanned downtime.

Rate of Interest of IMS

An IMS with predictive maintenance capabilities will autonomously alert relevant personnel about issues and only recommend maintenance activities when necessary. In this way, maintenance efficiency is significantly improved, offering manufacturers a number of benefits that positively affect the bottom line:

- Lower maintenance costs: Repairs are performed when needed instead of according to a predetermined schedule (which in many cases leads to redundant checks/maintenance activities).

- Increased uptime: With IMS, downtime can be scheduled, and as a result is usually much shorter than what's needed for reactive repairs.

- Reduced labor costs: Maintenance teams are smaller since tasks are planned beforehand.

- Lower equipment costs: Maintenance focuses on the problematic components only. This prevents wear and tear to surrounding parts during the repair and prevents unnecessary replacements.

- Lower inventory expenses: Since problems are predicted, orders are made only for materials and components that will be needed in the near future.

- Lower chance of secondary damage: Because problems are detected early on, they can be dealt with before more extensive damage is done to equipment.

- Increase in Remaining Useful Life (RUL): The root cause of issues is pinpointed making the need for disassembly less frequent.

- Quality 4.0: By improving asset health, deviations in performance become far less frequent, leading to consistently high levels of output quality.

Innovative Technologies Changing Maintenance Management in Oil and Gas Sector

Innovative and disruptive technologies are impacting every industry and way of life at a rapid pace. While beneficial, it's fascinating to think that these innovations of today will be quickly outdated as they give way to or spawn other technologies.

Most of us are aware of current advancements in artificial intelligence, a technology that is on the threshold of becoming a major influence in the coming years. Self-driving motor vehicles have moved from a science fiction movie to an everyday reality that is set to permanently change our view of transportation.

A technological advancement in one industry can be imported and applied to a completely different sector. For example, developments in materials and designs for space programs have been adapted for firefighters to create lightweight breathing apparatus.

Maintenance related activities are not often associated with innovation and disruptive technologies but the reality is that there is often pressure to reduce costs and downtime. This impetus drives the change which has a direct impact on maintenance management. Early adopters of technological breakthroughs can position themselves to reap the rewards of increased profitability and business viability.

Here are five innovative maintenance technologies that hold the potential to herald changes in the oil and gas sector:

1. A subsea robot arm called Eelume: Up until recently, undersea inspection and repair operations have been carried out by Remote Operating Vehicles (ROVs). A ship is used to carry the ROV to position, to be lowered into the sea and the inspection work is controlled by an operator on board the ship.

Enter the Eelume, a snake-like robot. The Eelume was developed for inspection, maintenance and repair of undersea oil and gas infrastructure and has caught a lot of attention from industry leaders. What separates it from other underwater repair technologies is the fact that this robot is self-propelled and therefore does not need the support of a remotely operated vehicle (ROV).

For this reason, the Eelume is considered a truly disruptive technology that will significantly reduce the costs of first-line inspection and maintenance of undersea installations. The Eelume has been developed by top academics from NTNU, Kongsburg Maritime and Satoil. Kongsberg Maritime has 25 years of experience in technology development with marine robots and Statoil provides access to real installations for testing and qualification.

According to the robot's developers, the primary cost factors in undersea maintenance are impacted by the increasing depth of new installations and the aging of the existing infrastructure. The Eelume can be stationed at the undersea facility and activated for maintenance activities as required, which will greatly improve the efficiency and reduce the time taken to execute a repair.

Typical tasks conducted by the Eelume will include visual inspections, cleaning and adjusting valves and chokes. Its snake-like shape will enable it to access hard to reach places between the network of pipes and infrastructure. Also, due to their constant presence on the ocean floor, the Eelume can be used to respond to emergency situations, mitigating a problem before it becomes a major repair.

2. The development of the Industrial Internet: Another advancement is the Industrial Internet of Things which utilizes the Internet of Things technologies in the manufacturing sector by combining machinery and "intelligent data", as described by the Industrial Internet Consortium. The consortium goes on to say, "The Industrial Internet will transform industry through intelligent, interconnected objects that dramatically improve performance, lower operating costs and increase reliability."

This sounds like every manufacturer's dream and the source of a significant change in the efficiency and effectiveness of maintenance related activity.

GE and BP are using this technology to launch an ambitious project to connect 650 deep sea oil wells to the Industrial Internet and then use GE's Predix Software Platform in order to monitor and predict the performance and failure of wells.

There are two things needed to optimize the maintenance and operation of complex equipment like that of a deep sea well:

- First, you need information: real-time access to process measurements that tell you what is happening on the well and the condition of the equipment.

- Second, you need high levels of technical expertise to interpret the data and make decisions based on maintenance required and predictions of future problems.

The benefit of this technology is that you can now develop analytical and predictive tools based on the data from 650 wells using the technical expertise of your best engineers – and it is online 24 hours a day. Think of it as cloning your best team of technical resources multiple times over to monitor and advise the maintenance crew in every production facility you have. It is estimated that an operator is set to lose up to $3 million per week if a well goes out of production. Technology that can, therefore, prevent an unplanned outage is extremely valuable.

3. Virtual Reality Immersive Training: One key component to keeping maintenance costs under control is training personnel and increasing their skills and effectiveness in making decisions. This is where virtual reality immersive training comes into play.

By using virtual reality technology, workers can get a realistic feel for the work environment they will encounter before they even get on location. They can be trained for specific tasks in advance, thus cutting down on valuable time on site and improving safety performance.

"An example of this is TOTAL's Pazflor platform, for which the crew was able to begin its training while the facility was still under construction, "The training sessions in the virtual model reduced the time needed to prepare workers for their tasks, thus helping to put the oil platform off the coast of Angola into operation more than two months earlier than planned."

Virtual reality training companies like Systran are migrating gaming technology advances to the field of manufacturing and maintenance. For example, Systran's LOTO EXP Virtual Experience Trainer is being used to train workers on isolation and lockout procedures using a standard Xbox controller.

A grading system based on time taken and compliance to procedure enables workers to learn from their training and improve their readiness for onsite tasks.

4. Eddyfi inspection tool for corrosion under insulation: An advancement that is also likely to impact maintenance is a new technology for non-destructive testing (NDT) created by Eddyfi. In any oil or gas facility, the integrity of pipelines is critical for both production and safety. Any pipe rupture can have catastrophic consequences.

The purpose of NDT is to identify where corrosion or erosion is affecting the integrity of equipment (for example, a product pipeline). When this is identified, a planned maintenance project can be executed to repair the damage or replace the equipment before a failure results in a costly emergency repair.

When piping and vessels are covered by insulation, existing NDT technologies are limited in their effectiveness to measure pipe thinning. The situation is further complicated as a slight defect in the insulation can allow moisture to penetrate into the metal and accelerate the effects of corrosion and pipe thinning.

Eddyfi's program of NDT measurements will show pipe thinning over time, which will then prompt maintenance organizations to plan a repair or replacement before the thickness deteriorates beyond acceptable limits. When NDT results and predictions improve, it not only impacts the effectiveness of the planning and scheduling of repairs but also minimizes downtime.

Eddyfi has conducted in-depth studies of CUI (Corrosion Under Insulation) testing techniques and found Pulsed Eddy Current (PEC) technology to be the most effective under typical CUI conditions.

Their Lyft product is an example of an innovative technology that is adding value to maintenance crews. It predicts failures and allots time and space to effectively prepare complex maintenance tasks where insulated equipment is vulnerable to CUI, which is both high risk and often difficult to measure.

5. Fluke cabling installation and maintenance advances: Fiber optic cables are a standard infrastructure component in manufacturing facilities. These cables are utilized to transfer large volumes of data for process control, optimization and maintenance diagnosis. Timely availability of information is critical to many decisions required on a daily basis. Due to the hazardous nature of many manufacturing facilities, the number of resources required in the field and tasks that are needed to be done in hazardous environments are kept to a minimum. Up until now, the need to test a fiber optic system from both ends of the cable has demanded the deployment

of technical resources in the field for every new installation and maintenance check of an existing system.

Fluke Networks' OptiFiber Pro OTDR with Smartloop product addresses this problem. OptiFiber promises to reduce the test time for fiber by at least 50%. Using this new technology, technicians will no longer need to test a fiber link from both ends as a single test at one end can now evaluate links in both directions. This will also eliminate the need for technicians to travel to dangerous sites or areas difficult to access.

This is another perfect example of an innovative technology that decreases resource requirements and the time to perform a standard maintenance job, thus reducing costs associated with maintenance and project management.

Innovation Continues

Innovation and advances in technology that can be cross-transferred to other industries are creating opportunities for increases in efficiency and cost savings for maintenance crews.

Organizations stand to benefit from:

- Predictive technologies that facilitate proactive repair rather than reacting to emergency breakdowns.

- Optimized allocation of resources through technologies.

- Reducing time on tasks through new and innovative methods.

Technology is an ever-developing process. As advancements are made, they will quickly serve as stepping stones for future innovations.

Predictive Maintenance Industry 4.0

Predictive maintenance for industry 4.0 is a method of preventing asset failure by analyzing production data to identify patterns and predict issues before they happen. Until now, factory managers and machine operators carried out scheduled maintenance and regularly repaired machine parts to prevent downtime. In addition to consuming unnecessary resources and driving productivity losses, half of all preventive maintenance activities are ineffective.

It is not a surprise therefore, that predictive maintenance has quickly emerged as a leading Industry 4.0 use case for manufacturers and asset managers. Implementing industrial IoT technologies to monitor asset health, optimize maintenance schedules and gaining real-time alerts to operational risks, allows manufacturers to lower service costs, maximize uptime, and improve production throughput.

How does IoT Predictive Maintenance Work?

For predictive maintenance to be carried out on an industrial asset, the following base components are required:

- Sensors: Data-collecting sensors installed in the physical product or machine.

- Data communication: The communication system that allows data to securely flow between the monitored asset and the central data store.

- Central data store: The central data hub in which asset data (from OT systems), and business data (from IT systems) are stored, processed and analyzed; either on-premise or on-cloud.

- Predictive analytics: Predictive analytics algorithms applied to the aggregated data to recognize patterns and generate insights in the form of dashboards and alerts.

- Root cause analysis: Data analysis tools used by maintenance and process engineers to investigate the insights and determine the corrective action to be performed.

Production asset data is streamed from the sensors to a central repository using industrial communication protocols and gateways. Business data from ERP and MES systems, together with manufacturing process flows, are integrated into the central data repository to provide context to the production asset data. Then, predictive analytics algorithms are applied to provide insights for reducing downtime, which are investigated using root cause analysis software.

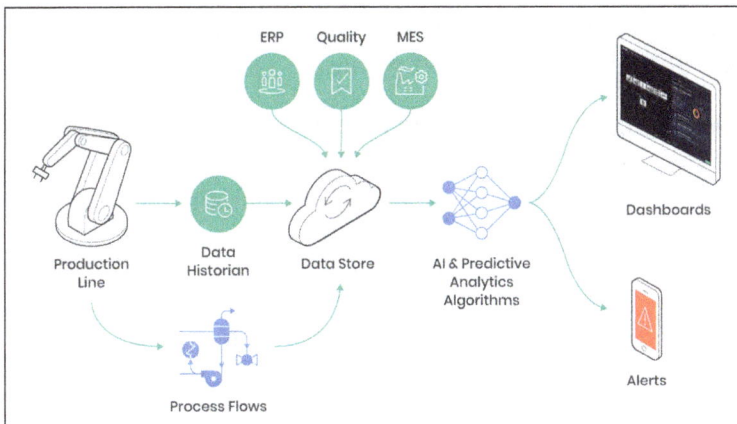

Predictive maintenance architecture.

To implement a predictive maintenance system effectively, manufacturers need to map the parameters of failure for machines and create a blueprint for their connected system – the manufacturing assets and sensors, business systems, communication protocols, gateways, cloud, predictive analytics, and visualization.

Using a visual IoT modeler, engineering teams can graphically capture the production processes in the shop floor, including data flows, dashboards, and the logic of the system – with rules that monitor and alert to maintenance issues. The modeler generates a system blueprint, which is critical for accurate predictive analytics.

Predictive analytics are applied to the machine data – and the system blueprint data – in order to predict conditions of upcoming failure. A dashboard for predictive analytics synthesizes operational data, allowing process and maintenance engineers to address actionable insights in the form of corrective action.

Benefits of Predictive Maintenance

Manufacturers and their customers get a range of business benefits from predictive maintenance. The advantages of PdM include:

1. Reduced maintenance time: Automatic reports for strategic maintenance scheduling and proactive repairs alone reduces maintenance time by 20–50 percent and decreases overall maintenance costs by 5–10 percent. These insights save the manufacturer and their customers time and money.

2. Increased efficiency: Analytics-driven insights improve OEE (overall equipment effectiveness) by reducing unnecessary maintenance, extend asset life and enable root cause analysis of a system to uncover issues ahead of failure.

3. New revenue streams: Manufacturers can monetize industrial predictive maintenance by offering analytics-driven services for their customers, including PdM dashboards, optimized maintenance schedules, or a technician dispatch service before parts need replacement. The ability to provide digital services to customers based on data presents an opportunity for recurring revenue streams and a new growth engine for companies.

4. Improved customer satisfaction: Send customers automated alerts when parts need to be replaced and suggest timely maintenance services to boost satisfaction and provide a greater measure of predictability.

5. Competitive advantage: Predictive maintenance strengthens company branding and value to customers, differentiating their products from the competition and allowing them to provide continuous benefit in-market.

Predictive Maintenance Tools

Implementing predictive maintenance requires a baseline of integrated tools. Predictive maintenance tools include an industrial IoT platform to model, simulate, test and deploy the predictive maintenance solution. The tools include industrial data integration and data analytics algorithms to detect patterns in machine data, and root cause analysis tools for investigating the derived insights and determining the corrective action to be taken.

Difference between Preventive and Predictive Maintenance

Manufacturers have been carrying out different forms of preventive and predictive maintenance for years. Understanding the difference between them, however, is critical with the emergence of Industry 4.0.

Preventive maintenance depends on visual inspections, followed by routine asset monitoring that provide limited, objective information about the condition of the machine or system. In this process, manufacturers regularly maintain and repair a machine to prevent failure. On the other hand, PdM is data-driven and relies on analytics insights for maintenance and repairs ahead of disruptions in production.

Companies using IoT Predictive Maintenance Tools

Organizations are implementing predictive maintenance analytics in a range of ways, from targeted solutions for a single machine part, to factory-wide deployments for increasing OEE throughout the production line. For machine and parts manufacturers, a relatively common predictive maintenance use case is monitoring and analyzing the condition of a motor to get alerts about its productivity levels, power consumption, health status, and internal wear.

Another powerful use case of predictive maintenance is minimizing production defects and reducing waste. Often referred to as Quality 4.0, such implementations can predict when the number of defective products is likely to exceed a threshold percentage, and provide the root causes for the expected failure.

Manufacturers are also turning to predictive maintenance for Factory 4.0, or a connected factory, by installing sensors in machines, workstations, and other designated sites such as the HVAC, security cameras or worker equipment, to predict issues across the factory floor.

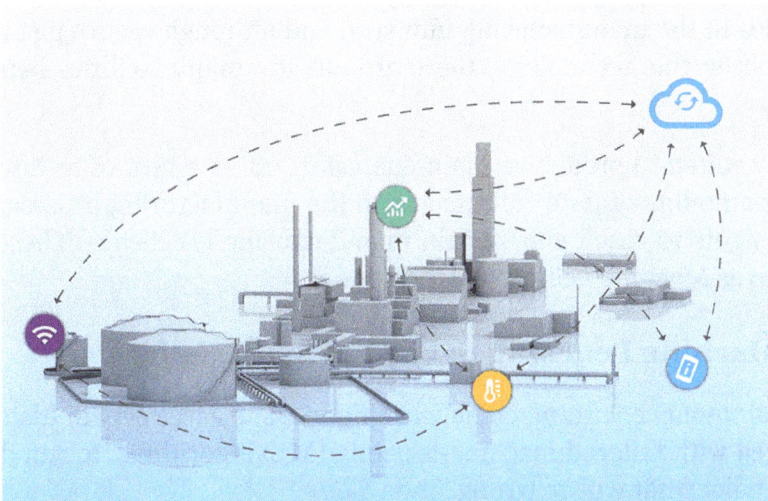

Common Approaches to IoT Predictive Maintenance

The two most common approaches to predictive maintenance are rule-based and machine learning-based:

Rule-based Predictive Maintenance

Also referred to as condition monitoring, rule-based predictive maintenance relies on sensors to continuously collect data about assets, and sends alerts according to pre-defined rules, including when a specified threshold has been reached. With rule-based analytics, product teams work alongside engineering and customer service departments to establish causes or contributing factors to their machines failing.

Once common reasons for product or part failures are established, manufacturers can build a virtual model of their connected system. Here they define product use cases, with "if-this-then-that" rules which describe the behaviors and inter-dependencies between the various IoT system components.

For example, if temperature and rotation speed are above certain predefined levels, the system will send an alert to an operator dashboard, to address the issue ahead of failure.

These rules provide a level of automated, predictive maintenance, but they are still dependent on a product team's understanding of what parts or environmental elements require measuring. The condition monitoring dashboards can be integrated with insight from machine learning to provide a visually understandable heat map of asset conditions in real-time.

Predictive Maintenance with AI

Industrial artificial intelligence can be applied to predictive maintenance and many other use cases in the manufacturing industry, and although we are just in the beginning of exploiting this technology, there are already many facilities benefiting from industrial AI.

AI is perfectly suited to predictive maintenance. It offers a host of techniques to analyze the huge amounts of data collected from the manufacturing process, and deliver actionable insights to reach and sustain manufacturing excellence. These techniques are referred to as Machine Learning algorithms.

Applying Machine Learning to Predict Asset Failure

Predictive maintenance with machine learning looks at large sets of historical or test data, combined with tailored machine-learning (ML) algorithms, to run different scenarios and predict what will go wrong, and when.

Predictive Maintenance ML Algorithms

Advanced AI algorithms learn a machine's normal data behavior and use this as a baseline to identify and alert to deviations in real-time. The algorithms required for machine learning must analyze input (historical or a training set of data) and output data (the desired result). A machine monitoring system includes input on a range of factors from temperature to pressure and engine speed. The output is the variable in question – a warning of a future system or part failure. The system will then be able to predict when a breakdown is likely to occur.

There are two main approaches to AI and machine learning for predictive analytics – supervised and unsupervised machine learning – each is relevant for a different scenario and depends on the availability of sufficient historical training data and the frequency of asset failure.

Predictive Maintenance Technologies

The start of predictive maintenance (PdM) may have been when a mechanic first put his ear to the handle of a screwdriver, touched the other end to a machine, and pronounced that it sounded like a bearing was going bad. We've come a long way since then with a variety of technologies for analyzing what's going on inside equipment, but the need for a knowledgeable, experienced person to use the technology hasn't changed. Today, as in the beginning, successful predictive maintenance is a combination of man and technology.

But let there be no mistake; it is the advances in technology that have made PdM a reality — most notably the ready availability of cheap computing power to gather, store, and analyze the data that makes PdM possible. By some counts, there are more than 40 technologies being used for predictive maintenance. Others might argue that many of these are simply variations on other, more basic methods. Presented here are some of the essentials regarding the most-used PdM technologies.

Program Concepts

Any predictive maintenance program should be characterized by a combination of three phases:

- Surveillance — monitoring machinery condition to detect incipient problems.

- Diagnosis — isolating the cause of the problem.

- Remedy — performing corrective action.

Consistent, accurate data gathering is essential to all three phases.

In general, the predictive maintenance process can be broken into 12 essential steps. The first six steps are actions to be taken before condition monitoring begins. Steps seven through ten represent the routine monitoring of an established program, and the final two steps cover diagnosis and fault correction.

Analysis of data is where the knowledge and experience of maintenance personnel becomes most important in a PdM program. It normally requires extensive training not only in the analysis techniques, but also in the use of the particular hardware and software employed. There are five important analysis techniques in PdM:

- Data comparison: Recognition of changes in data as compared to earlier data or baseline data on similar equipment.

- Limit or range tests: Specific testing to discover operating parameters that do not follow continuous trends or repeatable patterns.

- Pattern recognition: Identification of deviations from established patterns.

- Correlation analysis: Comparison of data from multiple sources, related technologies, or different analysts.

- Statistical process analysis: Use of statistical techniques to identify deviations from the norm.

Alignment

Misalignment of shafted equipment will not only cause equipment malfunctions or breakdowns, it may be an indicator of other problems. Checking and adjusting alignments used to be a very slow procedure. But the advent of laser alignment systems has reduced labor time by more than half and increased accuracy significantly.

Laser alignment systems for shafts have been available for many years. Laser devices for aligning sheaves and pulleys have recently come to market.

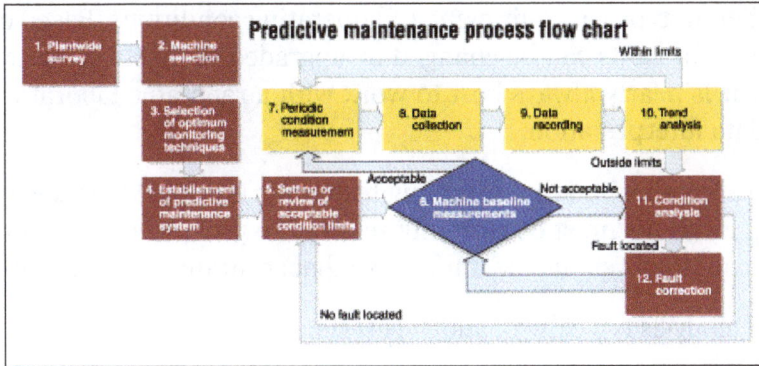

Predictive maintenance process flow chart

Although this program was developed for vibration analysis, it can also be applied to most other PdM processes.

Ultrasonic Testing

Instruments designed for ultrasonic testing sense ultrasound waves produced by operating machinery as well as the turbulent flow of leakage. They provide fast, accurate diagnosis of such problems as valves in blowby mode, faulty steam traps, and vacuum and pressure leaks. Ultrasonic observations may be taken in either airborne or contact mode.

Airborne ultrasonics is extremely useful in the location and diagnosis of mechanical problems, but the technology is not capable of isolating specific sources of ultrasound within a machine. Testing instruments are usually battery operated for portability. Their electronic circuitry converts a narrow band of ultrasound (between 20 and 100 kHz) into the audible range so that a user can recognize the qualitative sounds of operating equipment through headphones. Intensity of signal strength is also displayed on the instrument.

As scanners, ultrasonic instruments are most often used to detect gas pressure or vacuum leaks. Because they are sensitive only to ultrasound, they are not limited to a specific gas, as are most other leak detectors. In contact mode, a metal rod acts as a waveguide. When it touches a surface, it is stimulated by ultrasound on the opposite side of the surface. This technique is commonly used for locating turbulent flow or flow restrictions in piping.

Ultrasonic detectors are somewhat limited in their use. For example, they may help identify the presence of suspicious vibrations within a machine, but they are not sufficient for isolating the sources or causes of those vibrations. On the plus side, ultrasonic monitoring is easy (requiring minimal training), and the instruments are inexpensive.

Oil Analysis

Full benefit of oil analysis can be achieved only by taking frequent samples and trending the data for each machine in the program. The length of the sampling intervals varies with different types of equipment and operating conditions. Based on the results of the analyses, lubricants can be changed or upgraded to meet the specific operating requirements. It is nearly always best to work with a reputable laboratory for sample analysis and data interpretation.

It cannot be overemphasized that sampling technique is critical to meaningful oil analysis. Sampling locations must be carefully selected to provide a representative sample and sampling conditions should be uniform so that accurate comparisons can be made.

A thorough oil analysis typically includes 11 tests:

- Viscosity is one of the most important properties of a lubricating oil. The analysis consists of comparing a sample of oil from a machine to a sample of unused oil to determine if thinning or thickening of the oil has occurred during use.

- Contamination of oil by water or coolant can cause major problems. Because many of the additives in lubricants contain the same elements used in coolant additives, samples for analysis must be compared to samples of new oil.

- Fuel dilution of engine oil weakens the oil's film strength, sealing ability, and detergency. Dilution may indicate such problems as improper operation, fuel system leaks, ignition problems, improper timing, or other deficiencies.

- Solids content is a general test indicating total solids in the oil as a percentage of the sample volume or weight. Any unexpected rise in solids is cause for concern, because the presence of solids can significantly increase wear on lubricated parts.

- Fuel soot content is an important indicator for oil in diesel engines. Although fuel soot is always present in diesel engine oil to some extent, increases above normal levels may indicate fuel burning problems.

- Oxidation of lubricating oil can result in lacquer deposits, metal corrosion, or thickening of the oil.

- Nitration results from fuel combustion in engines. The products formed are highly acidic, and they may leave deposits in combustion areas and accelerate oil oxidation.

- Total acid number (TAN) is a measure of the amount of acid or acid-like materials in oil.

- Total base number (TBN) indicates an oil's ability to neutralize acidity. Low TBN is often an indicator that the wrong oil is being used for the application, intervals between oil changes are too long, oil has been overheated, or a high-sulfur fuel is being used.

- Particle count as part of a standard oil analysis is quite different from the wear particle analysis offered as a separate, specialized service. High particle counts indicate that machinery may be wearing abnormally or that failures could be caused by blocked orifices. Particle count tests are especially important in hydraulic systems.

- Spectrographic analysis reveals the presence of such elements as wear metals, contaminants, and additives in oil.

Wear Particle Analysis

While oil analysis provides information about the lubricant itself, wear particle analysis provides direct information about wearing conditions inside the machinery. This information is derived from the study of particle shapes, composition, sizes, and quantitites. Wear particle analysis is conducted in two stages. The first involves monitoring collected particles to determine normal conditions and trends. The second is the diagnosis of abnormal conditions as indicated by changes in the particle types, sizes, and quantities.

- Rubbing wear results from the normal sliding wear in a machine and should remain stable as a surface wears normally. But a dramatic increase in wear particles indicates impending trouble.

- Cutting wear particles are generated with one surface penetrates another, much as a cutting tool removes material. Cutting wear particles are abnormal and are always worthy of attention. These particles are produced when a misaligned or fractured hard surface produces an edge that cuts into a softer surface, or when abrasive contaminants become embedded in a surface and cut an opposing surface. Increasing quantities of longer particles signal a potentially imminent component failure.

- Rolling fatigue is associated primarily with rolling contact bearings and may produce three distinct particle types: fatigue spall particles, spherical particles, and laminar particles. Rolling spall particles are the most critical, because they indicate damage to a rolling element has already occurred.

- Combined rolling and sliding wear results from the moving contact of surfaces in gear systems. The chunkier particles result from tensile stresses on the gear surface, causing the fatigue cracks to spread deeper into the gear tooth before pitting. Scuffing of gears occurs when excessive heat from a high load or speed breaks down the lubricant film. Once started, scuffing usually affects each gear tooth.

- Severe sliding wear also results from excessive loads or heat in a gear system. Large particles break away from the wear surfaces. If conditions are not corrected, catastrophic wear is the likely result.

Infrared Thermography

Infrared thermography uses special instruments to detect, identify, and measure the heat energy objects radiate in proportion to their temperature and emissivity.

Midwave-range instruments detect infrared in the 2-to-5 micron range; longwave-range instruments detect the 8-to-14 micron range.

Infrared inspections can be qualitative or quantitative. Qualitative inspection concerns relative differences, hot and cold spots, and deviations from normal or expected temperatures. Quantitative inspection concerns accurate measurement of the temperature of the target.

As one of the most versatile predictive maintenance technologies available, infrared thermography is used to study everything from individual components of machinery to plant systems, roofs, and even entire buildings.

Infrared instruments include an optical system to collect radiant energy from the object and focus it, a detector to convert the focused energy pattern to an electrical signal, and an electronic system to amplify the detector output signal and process it into a form that can be displayed. Most instruments include the ability to produce an image that can be displayed and recorded. These thermographs, as the images are called, can be interpreted directly by the eye or analyzed by computer to produce additional detailed information. High-end systems can isolate readings for separate points, calculate average readings for a defined area, produce temperature traces along a line, and make isothermal images showing thermal contours. It is essential that infrared studies be conducted by technicians who are thoroughly trained in the operation of the equipment and interpretation of the imagery. Variables than can destroy the accuracy and repeatability of thermal data, for example, must be compensated for each time data is acquired. In addition, interpretation of infrared data requires extensive training and experience.

Vibration Monitoring

Vibration monitoring might be considered the "grandfather" of predictive maintenance, and it provides the foundation for most plants' PdM programs.

Monitoring the vibration from plant machinery can provide direct correlation between the mechanical condition and recorded vibration data of each machine in the plant. Used properly, it can identify specific degrading machine components or the failure mode of plant machinery before serious damage occurs.

Vibration monitoring and trending works on the premise that every machine has a naturally correct vibration signature. This signature can be measured when the machine is in good working order, and subsequent measurements can be compared with what is considered the norm. As the machine wears or ages, the vibration spectra change. Analyzing the changes identifies components that require further watching, repair, or replacement.

Most vibration-based PdM programs rely on one or more of the following techniques:

- Broadband trending provides a broadband or overall value that represents the total vibration of the machine at the specific measurement point where the data was acquired. It does not provide information on the individual frequency components or machine dynamics that created the measured value. Collected data is compared either to a baseline reading taken when the machine was new (or sometimes data from a new, duplicate machine) or to vibration severity charts to determine the relative condition of the machine.

- Narrowband trending monitors the total energy for a specific bandwidth of vibration frequencies and is thus more specific. Narrowband analysis utilizes frequencies that represent specific machine components or failure modes.

- Signature analysis provides visual representation of each frequency component generated by a machine. With appropriate training and experience, plant personnel can use vibration signatures to determine the specific maintenance required on the machine being studied.

Motor Analysis

Until fairly recently, predictive maintenance technologies for motors were limited to vibration testing, high-voltage surge testing for winding faults, meg-Ohm and high-potential tests for insulation resistance to ground, and voltage and current tests for testing phase balance. Many of these tests still have their place in plant maintenance, but several of them are impractical, dangerous, or harmful when tests are conducted with motors in place.

New technologies allow for portable, safe, and trendable tests that can be used for more accurate commissioning and troubleshooting. Each of these technologies has its strengths and weaknesses. But as part of a PdM program, they can accurately detect potential faults and avoid costly downtime.

- Static motor circuit analysis (MCA) provides a low-voltage, safe method of testing motor winding and rotor defects. The best instruments for this analysis use impedance-based tests coupled with insulation-to-ground testing. Impedance-based instruments are simple to use, and the results are easy to evaluate. Inductive-based instruments are for trending. Tests detect faults in motors, transformers, cabling, and connections. Motors must be de-energized.

- Motor current signature analysis (MCSA) is performed by taking current data and analyzing it using fourier transform analysis. Primary purpose of the test is rotor bar fault detection, but it is also useful for detecting rotor faults and power quality problems as well as other motor and load defects in later stages of failure. Motors must be energized and loaded during tests.

- Surge comparison testing uses high-voltage pulses to detect winding faults. Only experienced operators should conduct these tests because of the potentially harmful effects of high voltage impressed on used windings and cables. There are also challenges with testing assembled motors due to rotor effects on the motor circuit. Motor being tested must be de-energized with controls disconnected.

- High potential testing uses high-voltage ac or dc to detect faults to ground. Only the insulation condition between stator windings and ground can be evaluated and there is a potential for damage to the insulation system if the test is improperly applied or controlled. Motors must be de-energized with controls disconnected during tests.

Computerized Maintenance Management System

While the computerized maintenance management system (CMMS) was created in the 1960s as a punch-card system used to manage work orders, it's come a long way in the 50-some years since. Today, CMMS software is used to easily keep a centralized record of all assets and equipment that maintenance teams are responsible for, as well as schedule and track maintenance activities and keep a detailed record of the work they've performed.

The evolution of the CMMS to a cloud-based, multi-tenant solution has made it easier than ever for any maintenance team to harness the power of digital transformation.

Uses of CMMS Software

- Tracking work orders: Maintenance managers can select equipment with a problem, describe the problem, and assign a specific technician to do the work, either from a web or mobile app. When the machine is fixed, the responsible technician marks the work order "complete" and the manager gets notified that the work is done.

- Scheduling tasks: As a team starts to schedule preventive maintenance, they need a reliable work calendar. CMMS systems are especially good at scheduling recurring work and sending reminders to the right people. Organized maintenance scheduling helps even out the workload for a maintenance team making sure that tasks do not get forgotten.

- External work requests: Maintenance teams often have to take a work request from people outside the team. This can be a request from an assembly line operator who is hearing a strange noise from a drill or a tenant at an apartment building who is requesting shower repairs. The CMMS is a central place for recording these requests and tracking their completion.

- Recording asset history: Many maintenance teams have to care for assets that are 10, 20, even 30 years old. These machines have a long history of repairs. When a problem comes up, it is always useful to see how this problem was solved last time. In CMMS systems, when repairs are done, they are recorded in the machine's history log and can be viewed again by workers.

- Managing inventory: Maintenance teams have to store and manage a lot of inventory that include things like spare parts for machines and supplies like oil and grease. CMMS systems let the team see how many items are in storage, how many were used in repairs, and when new ones need to be ordered. Managing inventory helps control inventory related costs.

- Audit and certification: Many CMMS systems keep an unchangeable record of every action, so an asset's maintenance history can be audited. This is useful in case of an accident or insurance claim—an inspector can verify if the proper maintenance was completed on a machine. CMMS systems also keep data in a centralized system, which helps keep one version of the truth for ISO certification.

Where do you Install a CMMS?

There are two common places where CMMS software runs: On a computer at the client's business, or on the web:

On-premises CMMS

When a business is responsible for running its own CMMS, it is called an on-premises CMMS. The benefit of this kind of installation is the user has full control over network access to the CMMS server and complete data privacy (relevant for defense contractors, for example). The drawbacks are that this type of CMMS software implementation is expensive and complex. The IT department has to constantly look after the server, backups must be done by the maintenance team, and the software can quickly get out of date if updates aren't installed regularly.

Cloud-based CMMS

When the CMMS runs online, it is called cloud-based CMMS software. Here, the CMMS provider takes care of all the IT, hosting, security, and backups for the system, and the software can be accessed through any computer with an internet connection and web browser. Another strength of a cloud-based CMMS is that the software updates automatically, so you're always using the latest and most secure version. Make sure to subscribe to a cloud-based CMMS that lets you export the information in your current system so that you can migrate it if needed.

Benefits of a CMMS

- Measure maintenance performance: A CMMS makes it easy to do preventive maintenance, which means there are fewer surprise breakdowns and work outages. Allowing you to make better business decisions.

- Better accountability: Work order tracking makes it possible to quickly see if a technician did their work on time and get alerted when a task is complete.

- Less overtime: Better scheduling means that your team isn't sitting idle or working overtime, which means work can be distributed evenly.

- Information capture: Technicians can record problems and solutions, so this information is captured for others to use.

- Savings on purchases: Inventory planning features give you the time to shop around for spare parts pricing, instead of having to buy in a hurry.

- Certification and analysis: A full record of assets and performance helps managers analyze energy usage and plan maintenance spend.

Who uses the CMMS?

Every industry can benefit from maintenance care, and CMMS software can help businesses plan and manage that maintenance. There are four key user groups for these systems:

- Production maintenance: These are companies that make tangible products. They have machines, assembly lines, forklifts, and heavy equipment that require frequent maintenance.

- Facility maintenance: These are companies that take care of buildings. Apartment buildings, theatres, and government buildings all require maintenance. CMMS software helps them deal with structural, HVAC, and water-supply problems.

- Fleet maintenance: These are companies that take care of vehicles and transportation. Car rental companies, pizza delivery cars, city buses, transport ships, and fleets of towing trucks all need to have repairs scheduled, which can be taken care of with a CMMS.

- Linear asset maintenance: This is a special category of maintenance for companies that have assets like roads, water pipes, or fiber optic cables that cover great distances. A CMMS can help manage the complex maintenance required to keep these assets running.

What is the Best CMMS Software?

The best CMMS software companies are focusing on these areas for future development:

- Mobile CMMS applications: Maintenance workers spend most of their time outside the office fixing machines and taking care of buildings. So making the CMMS available in the field on their mobile phone is essential. With a mobile-friendly

CMMS app, technicians can record what they are doing as they are doing it, take pictures of the work, and request help onsite. Mobile apps that offer offline mode allow these updates to take place even without a Wi-Fi connection.

- Easy-to-use CMMS software: Many established CMMS companies make products that are very difficult to use. The interface hasn't changed since the late 1990's and many unnecessary, complicated features have been added to the product. More innovative CMMS companies often try to simplify the maintenance process and to make the software features easier to access and use.

- Fast CMMS data entry: The majority of CMMS projects that fail do so because they are too difficult to use and it is time consuming to enter data into the system. The next frontier for CMMS systems is focused on designing intuitive user experience and efficient ways to integrate data into the system from various sources.

- Cloud-based CMMS: New CMMS companies are mostly focused on providing a private CMMS for their clients, which runs online, through the cloud. The CMMS provider takes care of all the IT, security, and backups, making this a great option for modern maintenance teams.

Today's CMMS software helps businesses make informed decisions and stay on top of their work. A good CMMS system will streamline workflows and allow maintenance teams to easily keep a centralized record of all assets and equipment that maintenance teams are responsible for, as well as schedule and track maintenance activities and keep a detailed record of the work they've performed.

Good CMMS vendors are building for the future and are investing heavily in cloud-based solutions that are user-friendly and offer multi-tenancy, so that maintenance teams of any size at any number of locations can easily adopt the software. The evolution of mobile apps for maintenance teams is making it easier than ever for technicians to do work in the field, even when they're offline. It's important to select a CMMS that works best for your business, based on the features that are most important to your maintenance team.

Overseeing the maintenance of a facility, manufacturing operation or fleet can be a daunting task, to say the least. The introduction of computerized maintenance management software (CMMS) has not only changed the face of maintenance and facilities management, but has also improved the overall efficiency of maintenance departments both large and small.

Computerized maintenance management software enables the following 10 advantages:

- Plan and schedule preventative maintenance: Computerized maintenance management software automates the scheduling of inspections and maintenance, preventing the occurrence of maintenance problems and expensive repairs.

Switching from reactive to proactive maintenance also extends the life of equipment while reducing the overall operating costs of the maintenance function.

- Manage work orders efficiently: Software for maintenance and facilities management improves workflow and efficiency by allowing you to schedule, assign, and close work orders quickly and easily. CMMS software gives you the ability to configure work order screens with the desired fields, automatically tracks all work orders in the system, and captures the history associated with every piece of equipment.

- Manage spare parts inventory: When a maintenance department is disorganized, it can be difficult for maintenance managers to find the parts they need, when they need them. A CMMS system allows maintenance managers to track assets that need to be maintained and set up the automatic reordering of parts, so they have the right spare parts on hand and can perform repairs quickly. A CMMS also tells you exactly where the parts you need are stored, so you don't waste any time searching through storerooms.

- Eliminate paperwork: Using software for maintenance and facilities management eliminates the need for paperwork and clipboards because the software can be set up to capture information automatically. Furthermore, maintenance personnel can view all information related to work orders on their computers or mobile devices. Therefore, maintenance technicians don't have to search through folders and filing cabinets to find the information they need.

- Enhance productivity: Mobile CMMS can be used on smartphones, enabling maintenance technicians to access real-time information, check inventory, and initiate work orders without returning to the office. This reduces their journey time. Maintenance software also provides maintenance technicians with details about the procedures, parts, and tools necessary to perform a job, so they can work without delay or interruption.

- Reduce downtime and repair costs: Downtime is costly both in terms of revenue loss and damage to an organization's brand and reputation. When you focus on planned, preventative maintenance, equipment downtime is minimized. Because a CMMS system enables you to regularly maintain structures and assets, they are less prone to breaking down, which means that repair costs are also reduced.

- Increase safety: Computerized maintenance management software aids organizations in regularly checking and maintaining equipment and meeting safety standards to prevent malfunction and critical failures. This minimizes the loss of work time due to accidents and makes your equipment safer both for operators and the environment.

- Keep a finger on the pulse of your organization: Increase your understanding of your assets and organization by using the CMMS dashboard feature in your

CMMS to monitor key performance indicators (KPIs), look at the big picture, and create reports for decision makers. By using a CMMS to analyze historical data and trends, maintenance managers can identify problems areas, like rising costs, low productivity, or constant repairs.

- Ensure compliance with regulatory standards: Maintenance and facilities management systems must often comply with national and international regulatory standards. All maintenance managers face periodic audits or random inspections by regulatory agencies. Maintenance software allows you to demonstrate regulatory compliance and reduce the amount of preparation and paperwork that's required for an audit. Maintenance managers can simply generate reports detailing the maintenance work performed on vital machinery. This makes compliance easily traceable and reduces the risk of noncompliance penalties.

- Reduce overtime: Computerized maintenance management software can cut overtime significantly by reducing the need for emergency maintenance and repairs. By scheduling maintenance, maintenance staff can work more efficiently and effectively.

The cost savings that organizations achieve with the implementation of maintenance software can be staggering, but it's critical to choose a CMMS with the right functionality as well as work with the right partner to maximize your return on investment (ROI). An experienced vendor can provide you with the support you need today and as your maintenance operation evolves.

References

- Whats-so-smart-about-intelligent-maintenance-systems, datadriveninvestor: medium.com, Retrieved 11 January, 2019

- 5-innovative-technologies-changing-maintenance-management-in-the-oil-and-gas-sector, posts: prometheusgroup.com, Retrieved 14 July, 2019

- Predictive-maintenance: seebo.com, Retrieved 18 April, 2019

- Predictive-maintenance-technologies: plantengineering.com, Retrieved 7 May, 2019

- 10-advantages-of-using-computerized-maintenance-management-software: dpsi.com, Retrieved 13 July, 2019

Permissions

All chapters in this book are published with permission under the Creative Commons Attribution Share Alike License or equivalent. Every chapter published in this book has been scrutinized by our experts. Their significance has been extensively debated. The topics covered herein carry significant information for a comprehensive understanding. They may even be implemented as practical applications or may be referred to as a beginning point for further studies.

We would like to thank the editorial team for lending their expertise to make the book truly unique. They have played a crucial role in the development of this book. Without their invaluable contributions this book wouldn't have been possible. They have made vital efforts to compile up to date information on the varied aspects of this subject to make this book a valuable addition to the collection of many professionals and students.

This book was conceptualized with the vision of imparting up-to-date and integrated information in this field. To ensure the same, a matchless editorial board was set up. Every individual on the board went through rigorous rounds of assessment to prove their worth. After which they invested a large part of their time researching and compiling the most relevant data for our readers.

The editorial board has been involved in producing this book since its inception. They have spent rigorous hours researching and exploring the diverse topics which have resulted in the successful publishing of this book. They have passed on their knowledge of decades through this book. To expedite this challenging task, the publisher supported the team at every step. A small team of assistant editors was also appointed to further simplify the editing procedure and attain best results for the readers.

Apart from the editorial board, the designing team has also invested a significant amount of their time in understanding the subject and creating the most relevant covers. They scrutinized every image to scout for the most suitable representation of the subject and create an appropriate cover for the book.

The publishing team has been an ardent support to the editorial, designing and production team. Their endless efforts to recruit the best for this project, has resulted in the accomplishment of this book. They are a veteran in the field of academics and their pool of knowledge is as vast as their experience in printing. Their expertise and guidance has proved useful at every step. Their uncompromising quality standards have made this book an exceptional effort. Their encouragement from time to time has been an inspiration for everyone.

The publisher and the editorial board hope that this book will prove to be a valuable piece of knowledge for students, practitioners and scholars across the globe.

Index

www.ingramcontent.com/pod-product-compliance
Lightning Source LLC
Chambersburg PA
CBHW061937190326
41458CB00009B/2758

* 9 7 8 1 6 4 1 7 2 6 5 7 3 *